W9-CIL-284

At the Heart of the Web

At the Heart of the Web

THE INEVITABLE GENESIS OF INTELLIGENT LIFE

GEORGE A. SEIELSTAD

HARCOURT BRACE JOVANOVICH, PUBLISHERS

Boston San Diego New York

Designed by Ezra C. Holston
Jacket design by Alex P. Mendoza
Chapter opening web by Deirdre Tanton

Printed in the United States of America

Library of Congress Cataloging-in-Publication Data
Seielstad, George A.
At the heart of the web.

Includes bibliographies.
1. Biology—Philosophy. 2. Life (Biology)
3. Human evolution—Philosophy. I. Title.
QH331.S432 1988 574'.01 88-21241
ISBN 0-15-139814-3

First edition
A B C D E

To Dolores
thence jointly to
Andrea
Carl
Mark

Man lies at the heart of a web, a web extending through the starry reaches of sidereal space, as well as backward into the dark realm of prehistory . . . It is a web no creature of earth has ever spun before. Like the orb spider, man lies at the heart of it, listening. Knowledge has given him the memory of earth's history beyond the time of his emergence . . . Even now, one can see him reaching forward into time with new machines . . . until elements of the shadowy future will also compose part of the invisible web he fingers.

Loren Eiseley

Contents

Foreword

Radio astronomers are lucky, at least those of us who live at our observatories. First, we get an opportunity to study the only universe we could study. That is to say, the universe needs to be almost exactly the way it is, and had to be almost exactly the way it was, if we, the living, were ever to be part of it. A considerable effort will be made in this book to discuss the narrow range of circumstances through which events navigated to provide a setting suitable for us.

The second reason I consider us lucky is that the universe revealed by radio waves emphasizes distance more than the one revealed by waves of visible light. The brightest few hundred radio objects in the sky, with the exception of a mere handful, are entire galactic systems at distances reckoned in billions of light-years; the brightest few hundred visible objects, on the other hand—again with the exception of only a handful—are stars within a few thousand light-years of the sun. Because signals from distant objects arrive only after long transit times, radio astronomy's vaster horizons afford a grander sweep of time. Traveling through that tunnel into the past, we see a universe in constant flux; we are aware that the universe is not frozen in the state it occupies today. In fact, the moment "now" is only an intersection along the temporal axis between time domains we call past and future. Such a coordinate system is centered on us and, therefore, is not an appropriate frame for dating cosmic events. We ourselves are merely temporary transmitters of that which we inherit to those to whom we bequeath it.

Another example of radio astronomers' luck is that our observatories must be located far from urban hubbub. Otherwise, the only part of the universe we could study would be our immediate neighborhood, with its din of radio frequency interference. Because most radio observatories are in rural settings whose splendor is inescapable, we retain intimate contact with the natural world. I am sure this influences many of us, including myself, toward believing that life is the most arresting feature of the universe. Had I lived, as a majority of my fellow citizens do, in an environment where nature was not permitted to intrude, my outlook would certainly be different and, I suspect, more limited. Living always where the natural world surrounds one teaches that nature is not "out there" for us to view as we would a scene on television. Instead, we are part of it, and it is part of us. Our kinship with the rest of the living world is manifest, as is our bond with the inanimate world, because parts of it have been or will be incorporated into the living.

In the end, the most revolutionary influence the radio astronomy profession exerts is via its extension of the human senses. For millenia, people viewed the sky with eyes and, in the last few centuries, with telescopes, sensitive only to a minute portion of the full electromagnetic spectrum. With characteristic human naivete and hubris, they formulated models of the universe that seemed, to their severely filtered vision, thorough in their sweep. But radio astronomy, by opening a new window on the universe, exposed components whose existence had not even been suspected. One message became clear: Observers matter. We are participators in defining reality. We must therefore guard against limiting our paradigms for existence because of limitations in our sampling of it.

One concludes that intelligence is a factor in determining what the universe is because only with it have humans been able to overcome such limitations as a vision confined to a sliver of the totality of cosmic electromagnetic energy. Humans use the tools they create—of which radio telescopes are outstanding examples—to give the universe an awareness of itself. But since we thereby become sensory apparatus for a system of of which we are an integral part, we also gaze upon ourselves. Are we parts of an inescapable loop that feeds back the information we measure to affect what we have measured? Do we change the universe by our attempts to quantify it?

Such questions are both grand and unavoidable. I have tried to provide answers that are informed, but confess that they may not be definitive. Our explorations of the universe are too primitive to ascribe definiteness to them, but only by seeking answers do we fully exploit our creative potential. Readers are invited to address the questions introduced, rather than passively absorb the message the author delivers. This treatise is a provocation, not a sermon.

That a treatise exists at all is a tribute to several contributors. I wish to thank Betsy Dyer, Jeremiah J. Lyons, N. H. Horowitz, George Field, William J. Kaufmann, and a publisher's referee who prefers to remain anonymous for their helpful suggestions for improving the manuscript. The editors at Harcourt Brace Jovanovich—particularly Klaus Peters, Susan J. McCulley, Amy Strong, and Ezra C. Holston—were especially helpful as well.

Edward O. Wilson, whom I have never had the pleasure of meeting personally, also had a role in assuring completion of this book. All writing endeavors, I suspect, proceed in fits and starts. During one of this book's "fits," I read Wilson's *Biophilia*. The resonance with nature he so eloquently and movingly expressed inspired me to spell out my own vision of humanity's place in nature, including its cosmological component.

Beatrice Sheets and Sue Shears labored diligently to convert the words and thoughts that emerged from my mind into a readable format. I am grateful to them, as well as to Ron Monk, who produced photographic prints of every illustration, often quickly. My indebtedness is greatest, albeit anticipatory, to the readers.

George A. Seielstad
Green Bank, West Virginia

ONE

Dawn of a New Era

We have come up from the caves; predatory and primitive ages drift behind us. With almost the suddenness of a nova's burst to glory we have entered a new dimension of thought and awareness of Nature. . . . We hold the future in a delicate and precarious grasp, as one might draw a shimmering ephemerid from the clutches of a web. . . .

Ansel Adams

HUMANS EMERGED, EVER SO RECENTLY, as natural *products* of an evolutionary process set in motion long before the planet they were to reside on had come into existence. Groping, as had all their relatives and ancestors—from ancient bacterium to recent ape—for a niche to call their own, an environment ideally suited to the advantages they possessed, an environment, in fact, that determined which of their capabilities were advantageous, humans quickly became *manipulators* of the very process that had spawned them. An advanced intelligence made even more powerful by pooling the contributions of all their kind, both contemporary and ancestral, granted them a degree of mastery over both their surroundings and the other organisms with which they shared them. Eventually, even extraordinarily rapidly in a cosmic sense, humans' collective intelligence positioned them at the dawn of a major new era: henceforth they could be *directors* of life's evolution.

That humans are poised to introduce a new era is illustrated by their recent space activities. One consistent characteristic of life, apparent since its emergence from inanimate matter and throughout its subsequent blossoming into today's intricate web, has been its relentless and creative thrust to exploit virgin niches. For example, some species pioneered routes to previously uninhabited oceanic islands, while even today others are quick to recolonize regions destroyed by natural catastrophes. Humans have extrapolated this exploitative thrust to another dimension by taking their first, still tentative steps toward niches beyond the bioplanet of their birth. A station in space, called *Mir* (Peace), was launched in February 1986 and has been continuously occupied ever since. Humans, at least a rotating handful of them, have created a permanently inhabited resi-

Fig. I-1. Plant life (near center) thrusts through recently laid lava in Volcanoes National Park, Hawaii.

Fig. I-2. Earth life at the farthest advance beyond the planet of its origin—Astronaut Edwin E. Aldrin, Jr., on lunar surface. (NASA.)

dence in space! They have left the cradle of their origin. One of their creations, the Pioneer 10 spacecraft, has gone even farther, beyond the solar system, in fact, on a course so remote its feeble electronic signals are no longer detectable on its place of origin. Will adventurous humans, many millennia hence, encounter other samples of living matter likewise lurking about the cosmos? Or is their transport of the living presence elsewhere without precedent in the lengthy history of the universe? Are we typical or unique? Are we taking a step that is inevitable during the blossoming of all bioplanets, or are we initiating an expansion of life into an otherwise sterile environment?

Another major question regarding the human presence within the universe concerns its inevitability. The evolutionary pathway that led to humans does appear, with hindsight, to have been traversed in just such a way as to guarantee their eventual introduction. It is as though that specific pathway had been built into the initial conditions of the universe. Subsequent chapters will detail how even slight changes, here and there, in any of numerous parameters would have obviated the human presence. But the appearance of inevitability can be deceptive. For had any other pathway been followed, resulting in some other outcome for the universe, the same *post hoc* reasoning would have concluded that the sequence of events preceding that outcome was ideal for producing it. The proof? It actually did produce it. Because it was proven "ideal" retrospectively, however, does not prove it was inevitable.

Fig. I-3. Model of Soviet Mir Space Station in Star City, USSR. An identical craft in Earth orbit has been inhabited by humans continuously since 1986.

Still, not all hypothetical universes seem equivalent. If some particular universe never contained thinking beings, then what would have reasoned as above to arrive at these conclusions? In other words, if nothing knew of a universe's past, or, for that matter its present, then what evidence would exist that there was a universe? Is a universe containing no awareness of its existence real? Or is it merely virtual or phantom? In the sense that our universe does contain beings able to observe, record, and contemplate its structure and history, it is perhaps special. So the conditions that fostered its special nature are significant.

This dilemma of accident or design can be approached in another way. Can we argue that, because life exists on earth, it was predetermined to inhabit a planet in just the right orbit about just the right star so that temperatures permitted water to exist in its liquid state? Most people would conclude that predestination was not involved. Instead, they deduce that conditions on earth were so amenable to the chemical reactions activating life, and other thermal environments so unfavorable, that life emerged in the only locale (or perhaps locales) where circumstances made it possible to do so. Life was not being clever or tactical. It was not doing anything at all. The universe was presenting an immense variety of physical and chemical environments, among which was at least one in which life would arise through the most natural circumstances.

Might we not reason similarly regarding our entire universe? The fact that it contains life, even though small changes in its history would have excluded it, does not mean life forced the universe to behave in a particular way. Life just awaited the existence of a universe with a history close to the particular one ours had. Perhaps other universes or other cycles of the present universe did follow lifeless space-time trajectories. If so, life arose purely by chance when—and probably as soon as—it was given the opportunity to do so. Events followed from the circumstances that preceded them; not the other way around.

The absence of a preplanned strategy in the universe's past

does not mean none will be present in its future. It will, and we humans—or, since the times involved are so vast, our evolutionary descendants—will be the architects of it. To assume otherwise is to condemn life, at least the solar system variety, to eventual extinction. For changes in the cosmos will occur in the future just as certainly as they have in the past. None will be noticeable in the span of a single human lifetime, since the cosmos marches to the pace of aeons, not decades; nevertheless, there will be change. For example, the life-support system into which we tap, namely, that fountain of energy, the sun, has a lifetime of some ten billion (10^{10}) years, of which half remain. Other stars likewise have finite lifetimes; none provides a suitable environment for eternity. Energy sources other than stellar ones, some possibly engineered, will therefore have to be exploited some billenia into the future. If, as seems certain, the energy density of the universe becomes ever more dilute, whole new survival strategies will be needed if life is to persist as far into the future as the universe itself has existed in the past.

But these challenges are remote, almost inconceivably so from a human perspective. If any kind of life deriving from our own

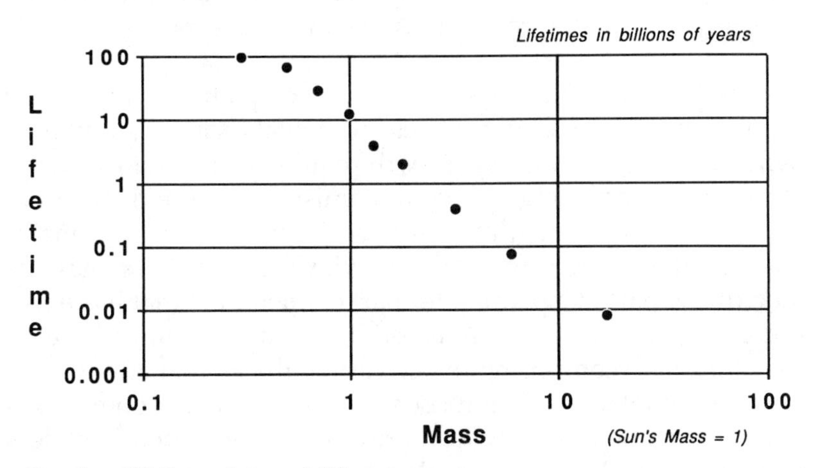

Fig. I-4. Lifetimes of stars of different masses.

is to remain to face them, it will only be because humans, acting *now* and with a sense of *urgency,* will have conquered their suicidal tendencies. They will have learned that their ability to manipulate nature does not license them to pursue only selfish objectives. Large-scale changes seemingly (albeit deceptively) beneficial to humans, but detrimental, or even deadly, to other inhabitants of the biosphere, will not have been pursued recklessly. For humans do lie at the heart of a web, a web so tightly woven that to sever one strand is to weaken the structure supporting all. If the web's time dimension does extend into the future, it will be because humans learned to regard life on earth as a *totality,* a single superorganism integrating several complex ecosystems into a whole that is greater than the sum of its parts. Each ecosystem will likewise be seen to be an integration of communities of diverse species into a *unit* exquisitely refined to exchange matter and energy with its surroundings; each community, a *synthesis* of individual organisms; each organism, an *integration* of cooperating organs; each organ, an *organized collection* of specialized cells; each cell. . . . With such a comprehensive understanding, humans will have learned the folly of promoting the cause of any one species, even if it is their own, at the expense of any other; to do so will seem no more sensible than to improve one's eyes at the expense of one's liver.

As with all life, the superorganismic bioplanet continuously changes as it ages and matures. Its metabolism, its repair of damaged "tissue," and its growth require a constant exchange of materials and energy. Such dynamism blurs the distinction, apparent at any single moment, between animate and inanimate matter. A carbon atom in today's atmosphere may tomorrow be part of a plant, later part of an animal that has eaten the plant, and finally be released again to the atmosphere by a bacterium decomposing the carcass of the animal after death. This perpetual turnover imposes another restriction on humans of the present era. If they wantonly abuse the presently lifeless matter—the soil, the oceans, the atmosphere—they ravage that

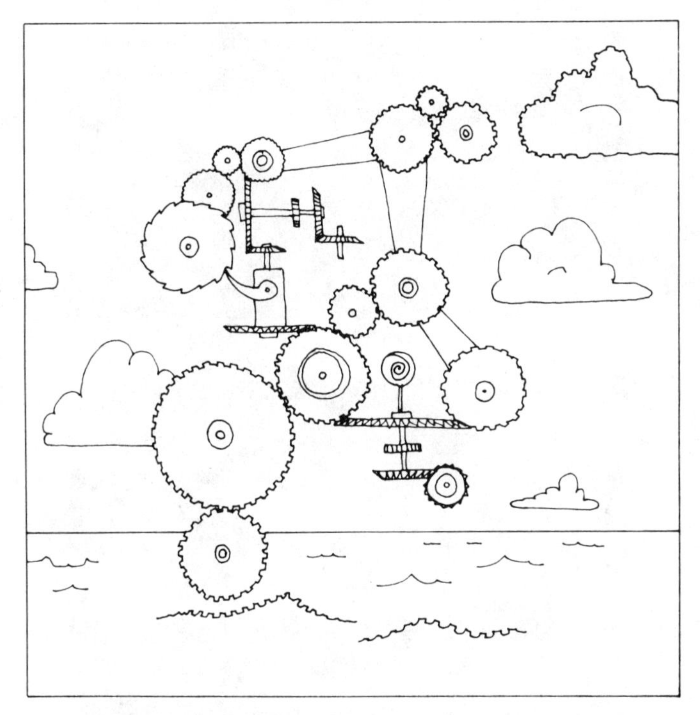

Fig. I-5. Life at all levels is as interconnected as the components of this mechanical contraption.

which will sustain the superorganism they reside within and thereby diminish the prospects for life in the future.

Underlying, perhaps driving, humans' advances toward their present epochal role in initiating a new form of evolution was the way each generation inherited an existing culture, expanded it, and transmitted the enhanced product to its successors. Skills and wisdom propagated to successive generations faster and in greater quantity than could ever have been the case had genes been the sole conveyors of information. What let this cultural evolution outpace biological evolution was the ability to communicate. Communications of all kinds—

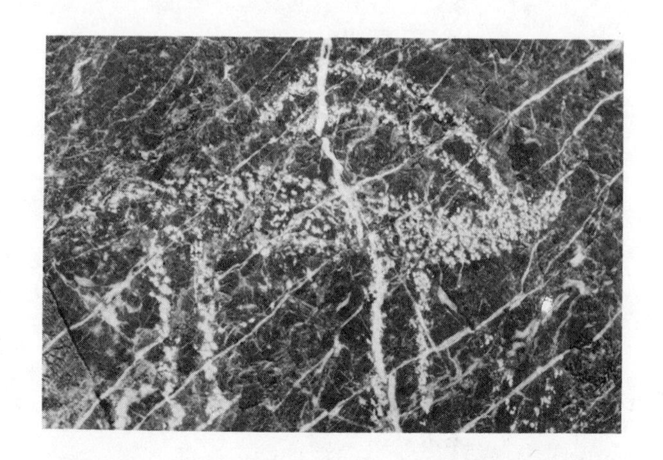

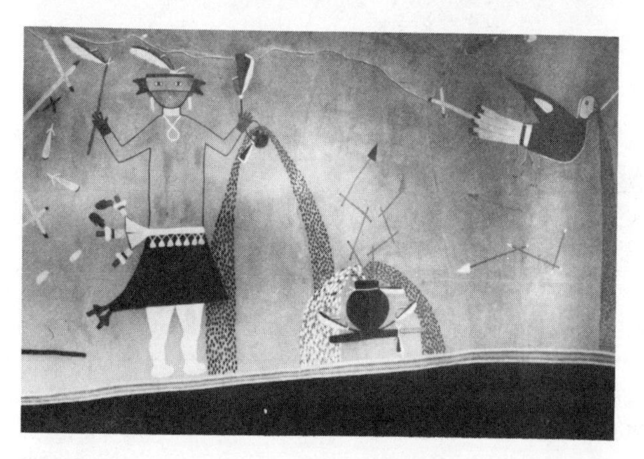

Seielstad

AT THE HEART OF THE WEB
The Inevitable Genesis of Intelligent Life

Errata

This page constitutes a continuation of the copyright page.

Credit lines for the introduction and chapter opening quotations are as follows:

Chapter 12: H.G. Wells, "The Discovery of the Future," *Nature,* #65, 1950, pp. 326–331.

Lines quoted on page 278 from the poem "The Double Axe" by Robinson Jeffers should have appeared as follows:

"... names
Foul in the mouthing. The human race is bound to defile,
I've often noticed it,
Whatever they can reach or name, they'd shit on the
morning star
If they could reach ..."

"A day will come when the earth
Will scratch herself and smile and rub of humanity ..."

verbal, symbolic, visual—wove a collective and cumulative human intelligence vaster and more powerful than that any individual or any single generation of individuals could possess. This pooled intelligence both created more useful information to communicate and improved abilities to do so. That is, communications among humans improved as they invented new technologies for that purpose (radio and television are modern examples), and technologies improved because greater quantities of information were being communicated. This interplay has a long history, dating at least to the time when a hominid ancestor's upright stance freed his hands for the construction of tools of wood, stone, and bone. Selection, both natural and cultural, thereafter favored those individuals and societies most skilled in communicating to successors their lore of craftsmanship, and these inheritors then advanced the crafts to new standards of perfection and utility.

As technology blossomed, it enabled humans to manipulate and modify their environment on an ever grander scale. Today, a brief tick of time on a cosmic clock since the stone tool era, humans have encountered some definite limits. Those of their tools capable of malevolent use, their weapons of warfare, have sufficient power to extinguish all life on the planet; no "improvement" in weaponry will enable greater devastation. A communication net has spread over the entire globe and is traveled at the speed of light, a limit that can never be exceeded. Never before has a medium conveying information, television, exerted near-simultaneous control over the thoughts and emotions of so large a segment of humanity. Individuals can address masses. One result has been the creation of an ideological chasm broader and deeper than any the world has ever seen

←

FIG. I-6. Humans communicate with each other and with their ancestors via an ever-increasing variety of techniques. (Top) Ancient Indian petroglyph in Titus Canyon, Death Valley National Monument. (Middle) Restored wall painting in Indian kiva, Coronado State Monument, New Mexico. (Bottom) Intelsat VI communications satellite. (COMSAT.)

Fig. I-7. Humans of this century have the power to extinguish all life on Earth. (U.S. Department of Energy.)

before. It divides the world into East and West, with each deeply distrustful of the other. Moreover, the speed of information transmission between East and West is swifter than either's social and political institutions can handle. Given the finality of the military option, the present generation of humans must immediately defuse the threat posed by mutual distrust if there is to be any living progression left to direct.

We humans, to whom evolution's future has been entrusted, must grasp from its past the message that random spontaneity produced the richness and complexity of life on this planet, of which we are only the most recent example. When crossing today's major evolutionary threshold, we must ensure that the process itself continues. Our newfound ability to engineer genetic material and our unrecognized responsibility to direct the course of evolution must not tempt us into arresting or slowing it. We should note that, had nature's replicative strategy, embodied in the DNA molecule, been perfect, our planet, if alive at all, would be populated by primitive, single-celled microbes only. Instead, DNA's fundamental instability punctuated exact reproductive succession with occasional mutations. Most resulted in failures, but a few launched evolution along new paths. Given that humans can, and therefore will, modify the genetic apparatus, our wisest strategy is to preserve ample fallibility. Cloning, remaking identical creatures, thwarts the opportunity for improvement. If tinker we must, let us follow nature's prescription by introducing greater variety.

One way in which humans can contribute in a beneficial way to life's development is by limiting their own numbers. If they do so, the social institutions that facilitate interactions among them—providing such services as education, health care, and entertainment, and distributing energy, resources, and wealth—will be able to keep up with the demand. These institutions are presently so overburdened and slow to respond that increasing the number of people they serve threatens their collapse. Moreover, restricting the numbers of humans in the biosphere can prevent the shrinking of variety in the gene pool, with all the lost opportunity that entails. Simply put, more people will utilize more resources; fewer will remain for other species; some will therefore be unable to survive; diversity will be sacrificed; and evolution will proceed with fewer options available.

Humans are obviously central characters in the pages that follow. But neither they nor any other species can be considered apart from their milieu. Each individual contains interacting units and is itself a unit in a larger interactive system. The boundaries separating units are indistinct. Even a superorganism the dimensions of bioplanet earth, for instance, is not self-contained; its metabolism is driven by energy from the sun, and its development has been significantly influenced by occasional cosmic events. The "environment" within which humans and their biotic companions exist is therefore unlimited, incorporating the whole of the cosmos. A major goal of this book is to develop this cosmological view of biology.

No multicellular creature springs into existence as a fully developed adult. Instead each starts as a single fertilized cell containing a full set of instructions for making duplicates, for assigning specialized tasks to them, for assembling them in such a way that the completed structure bears the stamp of its ancestry (that is, the union of two elephants produces an elephant), and for all other essentials in the growth and maturation process. These instructions differ only slightly for creatures as diverse as lizards and whales. All are "written" in the same "language" (that of the DNA molecule) using the same "alphabet" (a mere four nucleotides). Even the molecules essential to the formation of proteins, the amino acids, have the same "handedness" or sense of twist in all the earth's biota. Each lifeform is constructed of cells and employs the same method of cell duplication. The "energy currency" activating individual cells in all plants and animals is the same molecule. These similarities emphasize the underlying unity, the brotherhood, of all earth's life. They strongly suggest a common ancestry unfolding from a single, primitive, self-reproducing and metabolizing cell.

There is an analogy, then, between the development of the superorganism that is the bioplanet and that of an individual, complex organism, say a giraffe, although the superorganism's growth from unicell to maturity occurs at a drastically slower pace. It follows that a sample of the entire biosphere at any one particular instant does not adequately capture life's fullness, any more than a one-week study of a four-year-old giraffe summarizes that individual's total existence. Life has a time dimension as well as a spatial one. It has roots, knowledge of which is essential to comprehending it. So a second major theme of this book will be to weave life's presence into the temporal fabric of the universe. In pursuing this theme we shall discover that biology indeed has cosmological origins, that bio-

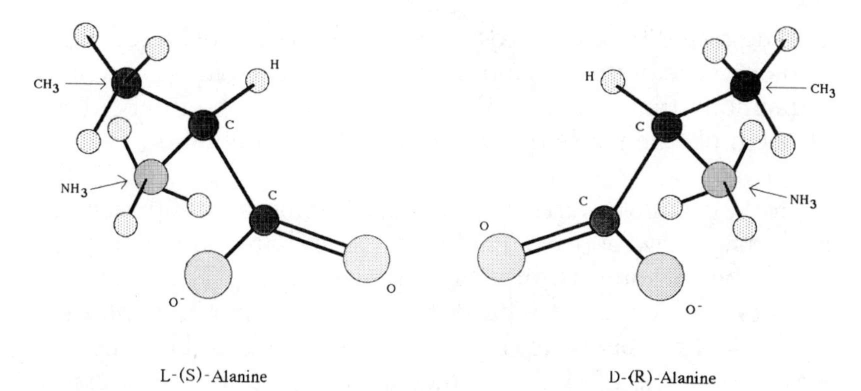

Fig. I-8. The amino acid alanine could assume either of two mirror-image forms, but only the L form exists in Earth's biota. (Drawing by Martha Forbes.)

planet earth is a progeny of earlier ordered structures that condensed, over time, from the chaos of the early universe.

Earlier we cited several factors testifying to the unity of all life on this planet. One senses that the similarities listed to support this unity were not necessarily unique. Perhaps they represented strategies that happened to work, so they were never changed. For example, one questions whether exactly twenty amino acids were necessary to serve as links in the macromolecules of which life is constituted. Could some have been excluded? Could other similar acids have been utilized? Could the macromolecules themselves have had the opposite handedness? Did life (considering here also its prebiotic ancestry) happen once upon a chemical foundation suitable for all its subsequent elaboration of form and function? Would the precise same foundation be selected if nature's grand experiment were repeated on earth? Would different biocommunities have emerged, also by chance, from the different sets of initial conditions certain to have prevailed on other planets? Specifically, would they have encoded their reproductive and assembly instructions in different molecular languages created from different alphabets? We shall hereafter adopt the catholic view of life suggested by these prospects. Life will be considered a natural outcome of universal principles. In that sense it is no more miraculous than the orbital revolution of one body about another in obedience to the universal law of gravitation.

History, cosmic or terrestrial, rarely marches across thresholds into major new epochs. The universe is doing so for only the fifth time in some twenty billion years or so. That immense span of time began at a particular instant when an explosion, now called the big bang, introduced all matter and energy into a point volume of space. Immediately, the universe began to expand, a process that continues to the present and will do so at least as far into the future, if not forever. But the total energy content—matter plus radiation—will not change.

When the first major epoch, called the Nuclear Epoch, began, the universe was so hot and dense that any particles of solid matter that spontaneously appeared out of pure energy just as

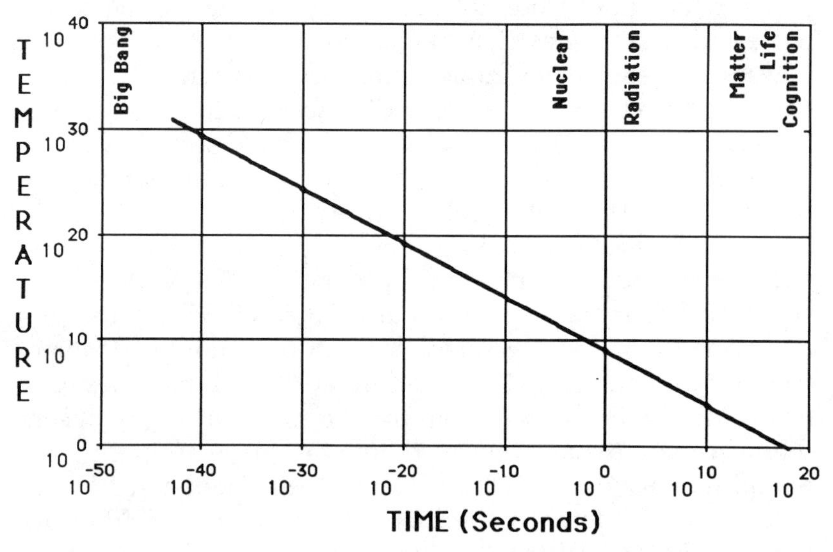

FIG. I-9. The thermal history of the universe, along which onsets of five epochs are marked.

FIG. I-10. Radiation becomes matter. A gamma ray photon (invisible) enters a bubble chamber (top), from an accelerator. Spontaneously it transforms into an electron-positron pair (electron curling clockwise, positron counterclockwise). A second pair created from another gamma ray originates near the center. (Lawrence Berkeley Laboratory, University of California.)

quickly disappeared by annihilating with their precise mirror-image antiparticles. The annihilating pair left behind—again—puffs of pure energy. Expansion, however, rapidly cooled the energy and dispersed the particles. Eventually (in reality a matter of only seconds or less—events moved rapidly in the early universe) some of the newly created particles were able to escape annihilation. Within minutes, the heavy ones—protons and neutrons in particular—began to fuse into bound units still heavier. As a result, the universe became populated with nuclei of hydrogen, deuterium (so-called heavy hydrogen), helium 3,

TABLE I-1. Major Heavy Elements in Nuclear Epoch

Element	Symbol	Protons	Neutrons
Hydrogen	H	1	0
Deuterium	^{2}H	1	1
Helium 3	^{3}He	2	1
Helium 4	^{4}He	2	2

and helium 4 (ordinary helium). Nuclei containing more protons than ^{4}He, and after less than 10 minutes even ^{4}He itself, could not be forged: the temperature of the universe was falling rapidly just at a time when multiply charged nuclei, to fuse, needed more energy to overcome their stronger electrical repulsion against like positive charges. Hence the Nuclear Epoch ceased only minutes after it had begun.

The second epoch, the Radiation Epoch, was long but boring. Bare nuclei and free electrons coexisted in a sea of radiant energy (colloquially light or, more formally, electromagnetic radiation). The density of the universe was still so high that packets of light energy (called photons) and electrons bounced off each other. Photons consequently cannot be traced directly to their origin; we cannot "see" where they came from, any more than we can see where sunlight originated near the sun's center for the very same reason of multiple scatterings off free electrons. The early universe was therefore opaque. Furthermore, light—for a good fraction of a billion years—was sufficiently intense and energetic to prevent electrically neutral atoms from forming; any that did were almost immediately separated again into their constituent nuclei and electrons the very next time they encountered radiation. Only after perhaps 700,000 years did the universe cool enough so that neutral atoms could persist, ushering in another major epoch of the universe, the Matter Epoch.

During this third epoch great assemblages formed. Radiation

could no longer disrupt the motions of particles of matter. Gravity was the strongest force to which they responded. Each felt the tug of every other, and, whenever some chance fluctuation created a density in excess of the average, that fluctuation became anomalously attractive. Its above average gravity pulled more particles into it, adding even more to what became a gravitational sink. In this way galaxies, stars, and planets—the prominent structures in the firmament—formed during the Matter Epoch.

The formation of a star is no ordinary event. So strong is the gravitational attraction among the atoms within one that conditions not seen since the Nuclear Epoch of the universe reappear at the star's center. There, densities and temperatures, driven by compression, rise to levels at which thermonuclear fusion can once more take place. First and longest in a star's life, hydrogen fuses into helium. When that process ceases for lack of hydrogen, helium fuses into carbon, then carbon into oxygen and nitrogen . . . on and on up the nuclear ladder, each stage briefer than the last, until eventually gravity can no longer squeeze matter to the temperatures necessary to fuse massive nuclei together. The degree of trauma this cessation of fusion introduces into a star's life depends on how massive it is. The usual outcome is for the core to collapse inward as the periphery is blown outward, the speed and violence of these dissociations increasing with the mass of the star. Note the general result: the matter emerging from a star when its fusion furnaces are permanently quenched has a greater chemical variety than the matter that entered it during formation.

Chemical evolution via stellar element synthesis continues to the present. In that sense the Matter Epoch persists. But raw atoms of whatever variety are not its sole or final product. Atoms have electrical affinities for each other and, when they can be brought closely enough together, often unite into molecules. Dense, cold, and dark regions of galaxies were suitable sites for these activities. They also were places where new stars could form, and with them planets, which, by virtue of their place of origin, enjoyed bounteous molecular inventories. On one of these planets, perhaps on more, chemical reactions among small molecules bonded them into larger ones. Innumerable combinations arose. Some of the resultant large molecules (or polymers) promoted specialized reactions by acting as templates for their own replication. Once such replicating chemicals came into existence they quickly outnumbered all others; for the replicators actively assembled themselves while the others passively awaited chance formation. The replicating reactions would have been especially efficient where the necessary assembly units were concentrated. There therefore existed a selection pressure favoring chemical reactions within encasing membranes. At what point a self-replicating, mutable system, capable of constant change, became a living entity is debatable: the boundary between pure chemistry and minimal biology is indistinct. But at some moment rather early in the history of our planet, the universe entered a new epoch, the fourth, the Life Epoch.

The Life Epoch has persisted for close to four billion years (on this planet, perhaps longer elsewhere), but it has been far from static. Rather, from its inception life has proliferated and diversified. At long last it has progressed to the moment—the cosmic Now—when it provides the universe with an awareness of itself. This inaugurates the fifth epoch, the Epoch of Cognition. Knowledge has long been associated with power; when knowledge advances to the point at which it can trace its own development, the power it conveys ascends to a wholly new level. Rather than passively witnessing the evolution that has characterized cosmic history up to this Epoch, the wielders of the new power bestowed by universal knowledge—namely, we humans—now have the capability to guide and shape that evolution into a cosmic future. So profound is this change that few are aware it has occurred. Each of us may be analogous to a single cell in a human body; just as that cell is not aware of the consciousness possessed by the total system to which it contributes, an individual human may not be aware of the cosmic consciousness possessed by the totality of his species. Nevertheless, those of us alive at this moment of unique cosmic significance must sense its epochal magnitude; for we can either help life continue its diversification and proliferation—into the niche of space, say—or we can extinguish it.

Only at these rare moments of severest challenge are opportunities for greatness granted. The magnitude of the challenge of preserving a continuity of earth life into the universe's future is unprecedented. It follows that the opportunities for greatness are magnified correspondingly. Readers of this book are invited to seize the opportunity.

For Further Reading

Grobstein, Clifford. 1974. *The Strategy of Life.* 2nd edition. San Francisco: W. H. Freeman.

Margulis, Lynn, and Sagan, Dorion. May/June 1986. Strange fruit on the tree of life. *The Sciences* 26(3): 38–45.

Seielstad, George A. 1983. *Cosmic Ecology: The View from the Outside In.* Berkeley: University of California Press.

Shklovskii, I. S., and Sagan, C. 1966. *Intelligent Life in the Universe,* pp. 182–200. New York: Dell Publishing Co., A Delta Book.

Thomas, Lewis. 1979. *The Medusa and the Snail,* pp. 27–30, 51–56. New York: The Viking Press.

White, Frank, 1987. *The Overview Effect.* Boston: Houghton Mifflin Company.

TWO

A Universe Fit to Live In

The tiny bit of space I occupy is so minute in comparison with the rest of the universe, where I am not and which is not concerned with me; and the period of time in which it is my lot to live is so infinitesimal compared with the eternity in which I have not been and shall not be. . . . And yet here, in this atom which is myself, in this mathematical point, blood circulates, the brain operates and aspires to something too. . . . What a monstrous business! What futility!

Ivan Turgenev

IF ALL THE DIRECTLY DETECTABLE MATTER in the universe—the stars, the galaxies, the X-ray bursters, the radio quasars, the infrared "cirrus" clouds, in sum, whatever has been detected by any instrument—were smeared out uniformly over the whole of space, then every ten cubic meters of that space[1] would contain only a single hydrogen atom. In approximately every thirteenth such cube, a helium atom would substitute for the hydrogen atom. Chemical elements other than these two simplest ones are so rare as to be insignificant in a grand cosmic sense (although vitally significant to those of us who are made of them).

In addition to this matter whose presence is detected directly, there is a considerable quantity whose existence can only be inferred. Of course, the amount of the latter is difficult to estimate; it is, after all, invisible. Indications of its presence often come through a study of the motions of visible objects. These motions are governed by the force of gravitation, a force that depends directly upon the masses attracting each other. Even if, in a pair of objects, only one radiates an electromagnetic signal, the other is known to exist because it influences the path its companion follows. Likewise, in a system involving many objects, say for example, a cluster of galaxies, constraints on the motions of the individual member objects are exerted by the total amount of matter contained in the system. So a study of such motions reveals the presence of matter, whether or not it can be "seen" (i.e., detected with a telescope of any type). If the spread in velocities of the galaxies in a cluster is large,

[1] A cube whose volume is 10 meters^3 has all sides equal to 2.15 meters, or about 7 feet. The load in a large dump truck might approximate 10 m^3.

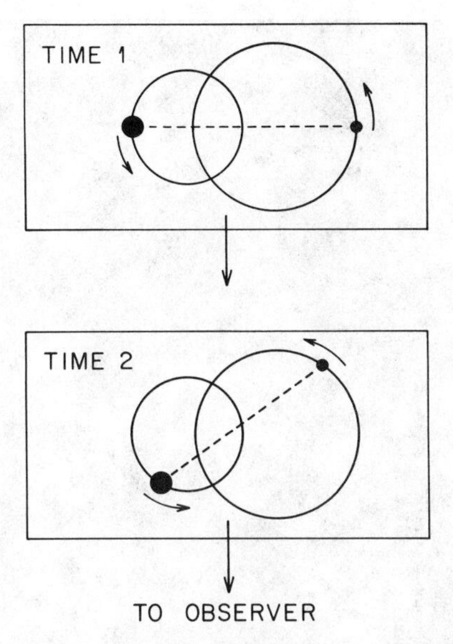

Fig. II-1. A pair of objects, each orbiting about their common center of mass. If only one radiated a signal to a distant observer, its companion could still be inferred from the existence of orbital motions in the detectable one.

the cluster itself must be massive; otherwise it would disperse as its members fled it. By the same reasoning, a small velocity spread among cluster members reveals a cluster with a small mass.

When the internal motions within gravitationally bound systems are analyzed in the way suggested, a surprise results: such systems seem to contain much more mass than can be seen. There is a good ten times more "dark" matter of unknown composition than there is visible matter. The overall density of matter in the universe, visible plus invisible, therefore approximates one atom—usually hydrogen, rarely helium—per cubic meter.

There is, however, more than matter in the universe. Radiant energy also fills the whole of space. In fact, every cubic meter contains approximately a billion (10^9) photons. Photons are

Fig. II-2. Coma Cluster of Galaxies. An example of a tight centrally concentrated cluster of more than a thousand galaxies. Their velocities spread over at least 930 kilometers per second, implying a mass in excess of a million billion suns (10^{15} solar masses), only a fraction of which is visible. (National Optical Astronomy Observatories.)

quanta of light, packets of electromagnetic energy. Each is massless and travels at the same speed, 300,000 kilometers per second (186,000 miles per second). The energy content of a collection of photons can be represented by a temperature. The warmth from an old iron poker that has been heated in a fire can be felt even before the poker's color has changed. A person who feels this heat is being bombarded by infrared photons. The distribution of photon energies radiating this warmth can be represented by a temperature of about 1,000 Kelvin degrees

(1,340° Fahrenheit).[2] If the poker is heated somewhat longer, it will glow dull red, then, later still, a "white-hot" color. The corresponding temperatures of the photon baths will be about 4,000°K (6,700°F) and 6,000°K (10,000°F), respectively. This last temperature is that assigned the photons in the visible light from the sun. Bluer stars, Vega for example, have hotter surface temperatures, 12,000°K (21,000°F) in this particular instance. In sum, cooler objects radiate photons of redder colors (longer wavelengths) than warmer ones. In addition, cooler objects radiate fewer total photons, regardless of their distribution among the various colors. This is another way of saying that low temperature correlates with less energy, as indeed one would expect. The billion photons that fill each cubic meter of space are cool indeed; their energy distribution reflects a temperature of just under 3°K (−454°F).

To even the most casual stargazer, this hypothetical division of the universe into small cubes, each containing a single light atom and a billion three-degree photons, must seem a totally inadequate description of reality. Where are the lumps—the planets, stars, and galaxies—that dominate his vista? The answer is that the lumps are insignificant compared to the amount of empty space. In other words, so much more of space approaches emptiness than contains matter that overall cosmic averages are little affected by the very occasional grittiness encountered. To illustrate the tenuous nature of the cosmos, consider, for comparison, a cubic meter of the air we breathe, a

---

[2] The Kelvin temperature scale starts at absolute zero, where the random dancing of a collection of particles ceases completely. It therefore has a more fundamental physical meaning than one based on phase changes of water, for example. Kelvin degrees have the same magnitude as Celsius degrees, both equaling 1.8 Fahrenheit degrees. The relations between Kelvin, Celsius, and Fahrenheit temperature scales are

$$°K = °C + 273.16 = (5/9)°F + 255.38.$$

As an example,

$$293.16°K = 20°C = 68°F.$$

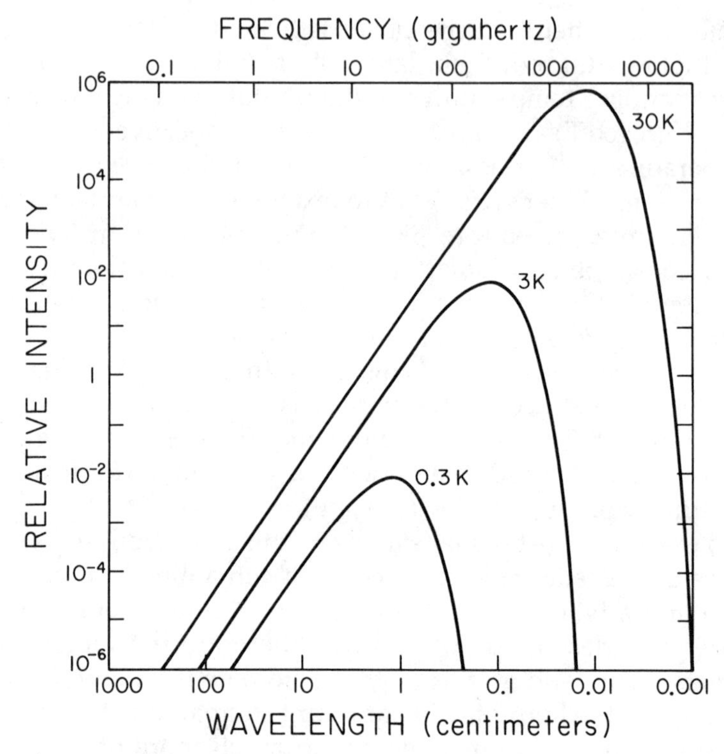

Fig. II-3. Energy distributions for perfect absorbers/radiators at various temperatures.

medium thin enough that we move through it with ease. Even so, each sample cube contains a few hundred trillion trillion (10^{26}, a one followed by twenty-six zeros) molecules. Sunlight adds an additional hundred trillion (10^{14}) photons, distributed in energy appropriately for a radiator at 6000° Kelvin (10,000° Fahrenheit); yet we are not torn asunder. The yawning emptiness of the universe is the first fact to absorb before studying the diverse structures that punctuate its vacuousness.

Of course, the reason for dealing first with grand cosmic averages is to assimilate the "big picture." We want to study the forest before considering individual trees. Can anything meaningful be deduced from the few parameters discussed so far? Indeed, yes. In fact, some of the most profound conclusions about cosmic history are derived from measurements of the radiant three-degree energy. For it, average properties suffice because it does uniformly flood the whole of space. What is the origin of this ubiquitous energy? Is it a remnant, a fossil, heat left over from the birth of the universe? Indeed it has been taken as such and offered as proof that space and time came into existence explosively. A fiery beginning bequeathed a legacy of heat—everywhere and for all time—because the spark of creation had no other space to which to escape. One must admit, however, that temperatures hovering close to absolute zero are not representative of fiery explosions; nor are low-energy radio photons what one expects to accompany a blinding flash. Fortunately, a mechanism for cooling gases and shifting photon energies is well known, namely, expansion of the volume in which the gases are confined. The present low temperature of the radiation filling the universe is consequently an indication of a second major property of our space-time framework, its continuous expansion.

Astronomers had deduced an expanding universe before the discovery of the 3°K background radiation, but the latter discovery clinched the case. The evidence previously involved the measured motions of galaxies. All but the nearest few appeared to be receding from our own, the Milky Way Galaxy. The velocities of recession increased with their distances from us. The interpretation was that each was being carried away, not only from the Milky Way but from every other galaxy, by the general expansion of the space all resided in. By running this scenario in reverse, astronomers concluded that at earlier times the universe was smaller and denser. When extrapolated backwards to its extreme, the universe's size became vanishingly small and its density infinite. Its temperature when in this compressed state was likewise unbounded. The residue of this heat energy is what now, as at the beginning, fills the entire universe, but expansion has reduced its present temperature to slightly less

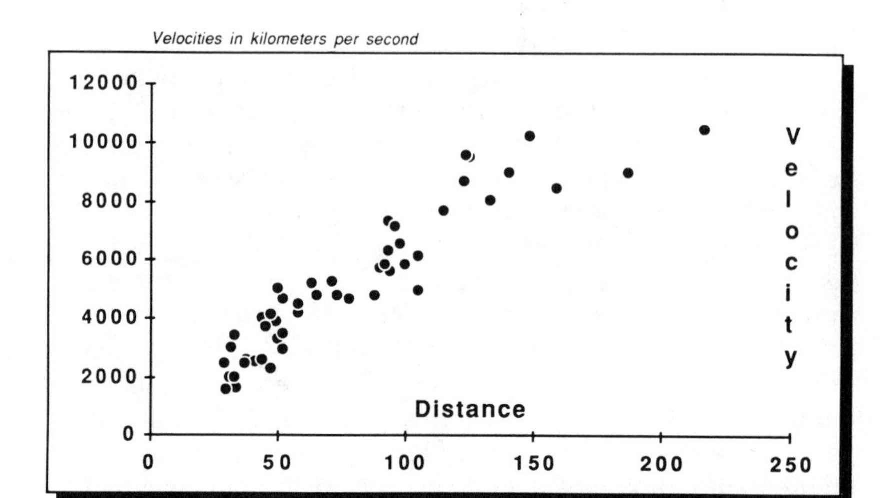

Fig. II-4. Relationship between velocity and distance for galaxies that are the brightest members of their clusters. (Data from M. Rowan-Robinson, The Cosmological Distance Ladder, *New York: W. H. Freeman (1985), Appendix A15, pp. 334, 335.)*

than three degrees above absolute zero. Because of this radiation's origin, we call it the "cosmic background radiation."

The mere presence of heat throughout the vastness of space, while necessary, is not sufficient to establish a primordial origin for it. If, for instance, discrete radiating objects—galaxies, quasars, clusters, whatever—fill space to enormous distances, then the likelihood that a telescopic search will eventually encounter a radiator, regardless of the direction of search, increases. Imagine, for the sake of analogy, standing within a forest of moderately spaced trees. If the grove has modest dimensions, one may peer through gaps between trees into fields beyond; but if the forest is extensive, every line of sight ends at a tree trunk. Are the mappers of the universe likewise screened from its most distant parts by a haze of intervening discrete objects? The prospect is unlikely because the three-degree fossil radiation is so prominent. It rises well above the measured and estimated elec-

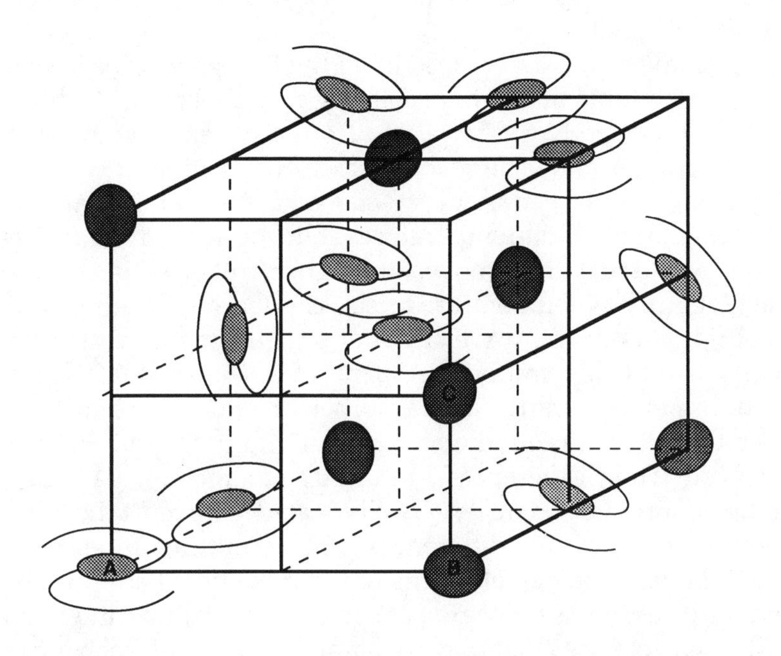

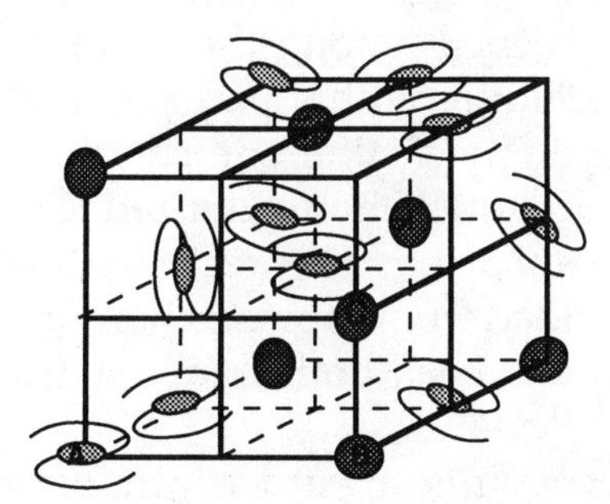

For every two units added to the separation of Galaxies A and B, one unit is added to the separation of Galaxies B and C.

Therefore, A separates from B at twice the velocity that C does.

FIG. II-5. *An expanding volume of space. As space grows, galaxies find themselves farther apart. The larger their separation, the faster they appear to be moving.*

tromagnetic contributions from a background haze of individual, radiating objects.

The shape of the curve describing the background of radiant energy also hints at a primordial origin. Well known to physicists, it is precisely that expected from a system that is both a perfect emitter and a perfect absorber of radiation, a so-called blackbody. The spectral energy curve has the same shape for perfect emitters of all temperatures, but the area under it rises rapidly, and its peak moves to higher frequencies (shorter wavelengths) as temperatures rise. The fact that the peak intensity occurs near a wavelength of one millimeter constitutes evidence that the emitter's temperature is approximately 3°K. To be a perfect emitter and absorber of radiation is to be in a state called thermodynamic equilibrium. Equilibrium is achieved when differences in various parameters, e.g., temperature, pressure, density, are eliminated. The most thorough method of eliminating differences is by permitting unrestricted interactions among all elements of the system. Where photons and particles of matter coexist at such concentrations and move with such velocities that they constantly collide with each other, a steady state is reached among them: for every photon that loses energy by colliding with a particle, another acquires the same amount through a different collision. Directions and speeds of motion are totally homogenized. Conditions favoring thermodynamic equilibrium are precisely those expected if the universe has expanded from previous hot, dense phases, when collisions would have been both frequent and violent, indeed inescapable.

Yet a third clue points toward a primordial origin for the three-degree background radiation: it arrives with equal strength from all directions in space; lumpiness in the form of localized spots that are either exceptionally hot or cold is absent.[3] A single word describes such a distribution, *isotropic*. No

[3] A small and smooth variation of temperature has been measured, but its cause is the peculiar motion of our own galaxy relative to a field of distant ones. Such motion is truly a minor and local perturbation on the overall cosmic expansion.

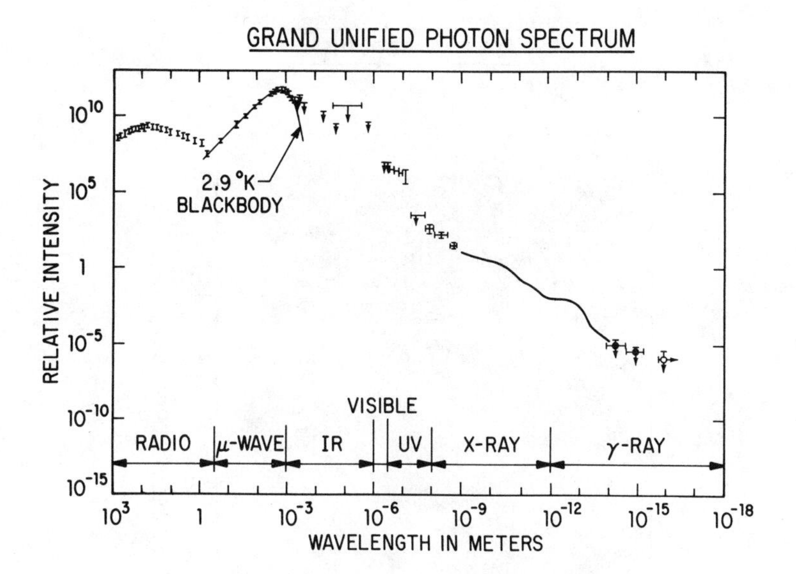

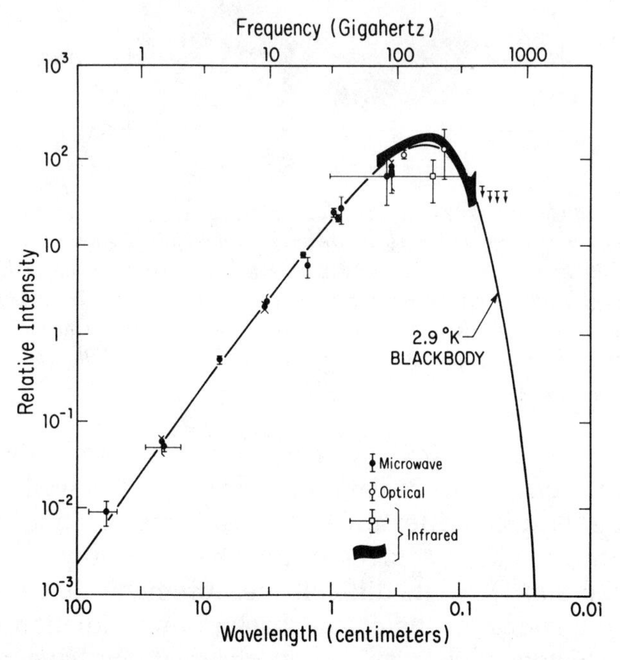

FIG. II-6. (Top) Background radiation across entire electromagnetic spectrum. The microwave portion, with its prominent peak labeled "2.9 K Blackbody," cannot be a synthesis of other radiating contributions on either side. (Bottom) Blowup of the 2.9 K Blackbody. The measured points are in nearly perfect agreement with the theoretically predicted curve for a system in thermal equilibrium.

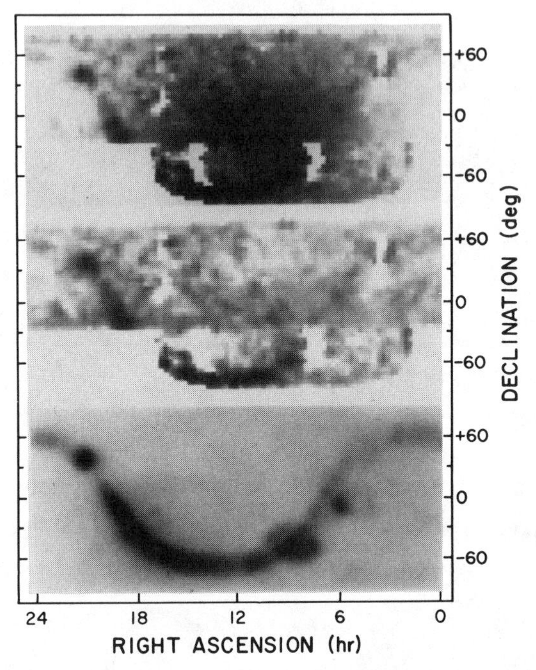

Fig. II-7. (Top) Map of sky at wavelength 1.2 centimeters. Gray scale for temperature, with darker representing hotter. Holes are unmapped regions. Temperature peaks near center, the direction toward which our galaxy is moving. (Middle) Smooth distribution of radiation remaining after temperature differences caused by our galaxy's motion are subtracted. (Bottom) Expected contribution from Milky Way Galaxy, showing it causes band in middle photograph. (Courtesy David T. Wilkinson, Princeton University.)

direction is special. Nor does any particular place stand out as an obvious origin; the radiation is *homogeneous* as well. It must, therefore, be a grand feature of the universe, not a local structure at all, for local sources cannot be so evenly distributed about the sun. Recent, difficult measurements have, in fact, placed the source of the cosmic background radiation beyond distant clusters of galaxies. X-ray observations have revealed that the space between the galaxies in these clusters is filled with a hot (some hundred millions of Kelvin degrees) gas of electrons. A few of the three-degree photons collide with these

TABLE II-1. Decrements in Cosmic Microwave Background's Temperature Caused by Intervening Galaxy Clusters

Cluster Name	Temperature Decrement[a]
0016 + 16	−0.0005 Kelvin
Abell 401	−0.0006 Kelvin
Abell 665	−0.0004 Kelvin
Abell 2218	−0.0003 Kelvin

[a] Decrease of sky temperature below about 3 Kelvin in directions to clusters.

cluster electrons on their way to our radio telescopes. The collisions scatter the photons to higher frequencies, producing a tiny dip in the temperatures measured in the directions of the clusters. Observations of these temperature decrements in the directions of clusters therefore indicate that the cosmic background radiation predated the formation of *any* concentrations of matter or energy in the universe. Its featureless spatial distribution is precisely what would result if the contents of the universe were stirred thoroughly together. This fossil radiation is the product of, as well as the best evidence for, a completely homogeneous and isotropic early universe. Furthermore, the initial expansion of the universe must have proceeded without regard for direction, since no departures from isotropy have evolved. We derive from a space remarkable for its absence of distinguishing landmarks.

Let us return to our simplified description of the universe to consider the significance of the ratio of helium to hydrogen atoms. The reader, a complex chemical assemblage living on a planet seemingly richly endowed chemically, may wonder, first, why so few kinds of matter are considered. The answer is that these two simple elements, one an order of magnitude more numerous than the other, account for all but a fraction of a percent of the (known) matter in the universe.[4] Each of us

[4] The nature and composition of the invisible matter are unknown.

represents a chemical anomaly, but even our summed contribution to the matter in the universe is much too insignificant to affect a cosmic average.

The existence of helium exemplifies one of the grand principles at work in the universe: large, complex units are formed out of smaller, simpler ones—stars out of atoms, galaxies out of stars, clusters out of galaxies, and so on. The fact that helium, with two protons and two neutrons in its stablest nucleus, formed from hydrogen, whose nucleus is a bare proton, is another indication that the universe passed through an extremely hot, dense phase. If it had not, individual neutrons and protons would never have been brought into the near contact necessary to bind them into identifiable nuclei. Separations as small as a few hundred-trillionths (10^{-14}, or 0.000,000,000,000,01) of a meter were required. These could occur only when the density of the universe was a million trillion trillion (10^{30}) times larger and the temperature a billion (10^{9}) degrees hotter than today. Under conditions that extreme, a helium nucleus represented a lower, more tightly bound energy system—therefore a more probable configuration—than a collection of four isolated nucleons.

The real triumph for those advocating a hot "big bang" origin of the universe comes not merely from their ability to explain the *existence* of ubiquitous helium, but also to explain its observed abundance relative to hydrogen. Using data acquired both with telescopes and with particle accelerators, they predict that about one-fourth of the material mass in the universe should be helium, the rest hydrogen.[5] Their predictions accord with observation. This synthesis of information concerning the largest and the smallest structures in the universe is one of the most exciting developments of modern science. It reveals the profound unity of the universe and the universality of physical

[5] Since one helium nucleus is four times more massive than a hydrogen nucleus, there must be twelve of the latter to every one of the former if the ratio of masses is to be H:He::3:1.

Fig. II-8. The radio telescope probes the depths of space (left), the alternating gradient synchrotron, the nuclei of atoms (right). Together they determine element formation in the early universe. (Courtesy National Radio Astronomy Observatory and Brookhaven National Laboratory.)

Table II-2. Measured Ratio of Helium 4 to Hydrogen

Object Type	Mass Ratio, ^{4}He/H
Solar System	0.26
Stellar Atmospheres	0.28
HII Regions	0.305
Globular Clusters	0.23
Planetary Nebulae	0.220
Blue Compact Galaxies	0.228
Low Luminosity Galaxies	0.245

laws derived to describe it. Insights like this are the treasures that motivate scientific research.

Why, though, were more complex nuclei not forged at the same time helium was being manufactured? The answer is that the universe was rapidly expanding, and its temperature dropping in synchrony, during this epoch of element formation. When the temperature fell below a few million degrees, at which time the universe was only minutes old, fusion of simple nuclei into complex ones ceased. So rapid was the initial rate of expansion that time adequate to manufacture a full complement of atomic nuclei was simply not available. Helium almost alone retains the imprint of the primeval universe. Like the three-degree background radiation, it is a cosmic archaeologist's delight.

Perhaps it is time to recapitulate. We have learned that the universe is filled with heat radiation. The spectral signature of this radiation—the way its intensity varies with frequency—is that emitted by systems in perfect equilibrium. No explanation for it other than the fossil heat left over from a hot, dense stage of the universe—when equilibrium could be established by the ceaseless tumbling together of the matter and radiation present—has ever been accepted. A tumbling epoch likewise explains why the radiation is both isotropic and homogeneous. Further evidence for a hot, dense stage is present in the relative abundances of the chemical elements. Helium, in particular, has exactly the abundance expected if its origin is primordial; if its origin is not primordial, then its observed abundance presently has no explanation.

We have also deduced a mechanism for cooling the radiation from the billion or so degrees required to forge nuclei to the present three degrees. That mechanism is expansion. All of space must be growing. As it does, the photons within it get farther apart; its radiation density is thereby diluted. This is another way of saying that the temperature of space cools. The story all hangs together: observations made now tell us a great deal about our past. But what about our future? Will expansion continue forever? Or will it gradually slow, perhaps eventually reversing into a contraction?

Whether the future of the universe is open without end or will occupy only some finite time span may influence our enthusiasm for acting in the interests of our descendants. Its resolution has an analogy in terrestrial rocketry. The ascent of all rockets is restrained by the gravitational pull the earth exerts. Nevertheless, if the initial launch velocity is large enough, a rocket can continue its voyage indefinitely—beyond even the solar system in the case of Pioneer 10, for example. Only if the launch velocity is low—less than seven miles per second from the earth—will the rocket's return be guaranteed. In the borderline case, the rocket's speed of separation from earth approaches ever closer to, but never precisely reaches, zero. Of course, the critical initial velocity depends on the mass and the size of the gravitating object. On the smaller and lighter moon, the critical escape velocity is only 1.5 miles per second, for instance.

The universe presents the same three options: continuing expansion, eventual reversal and subsequent contraction, or the borderline case of ever decelerating but indefinite expansion. Which future will ensue can be determined by comparing the rate of expansion with the mass density of the universe, that is, the outward momentum with the braking from within. We shall consider the quantitative details in a subsequent chapter. Here we want merely to mention the startling result: the actual density of matter, visible plus invisible, is close to the borderline value of five to ten hydrogen atoms per cubic meter just adequate to halt the cosmic expansion. Considering that the matter density could in principle be one ninety-fifth—or six hundred or any other number—times the critical amount, its nearness to criticality must have a cause. We shall provide one in Chapter IV.

What has any of this to do with the notion of a living universe? Merely this: life appeared when and perhaps because conditions made its existence possible. But conditions are not static phenomena; nor do they jump discontinuously from one state to the next.[6] Instead they evolve, or unfold, smoothly from the pattern that has preceded them. The fact that we (all living matter collectively) exist at all is because of our heritage. So we are part of a much vaster and more enduring continuum. That continuum transcends us: each of us will die, after the most ephemeral existence, but the universe will be ever renewed. Furthermore, whole species will go extinct, as, in fact, have the huge majority of those that have ever been present; species extinction is as normal as individual deaths. Our roots in time run deep, extending back to the very beginning of the universe. To appreciate this, note that many parameters of the universe's past, even some of those from its earliest moments, could not have deviated significantly from the values they in fact possessed if life were to be present someday.

For example, reconsider that hypothetical cube with its billion packets of energy (photons) for every single particle of matter. Obviously, only a tiny fraction of the energy present at the origin of the universe was distilled into the tangible form of particulate matter. But it could have been worse: no matter at all might have survived. It might all have vanished completely; in fact, it almost surely would have done so but for a slight asymmetry present in the early moments of the universe.

Ever since Einstein pointed out that matter and radiation are different manifestations of the same entity, energy, scientists have been aware of the possibility of interchangeability between the two (see Figure I–10). If a photon contains sufficient energy, it can spontaneously transform into a pair of particles. The particles are nearly, but not quite, identical: their electrical charges have opposite signs. One member of the pair is designated ordinary matter, the kind we—and, in fact, the whole earth, the galaxy, and as far as we know, the universe—are

[6] For the moment we are ignoring quantum or microscopic phenomena in favor of macroscopic ones.

made of. Its mirror image is called antimatter. Thus the familiar electron, with its tiny mass and negative unit of electric charge, is matter; its antimatter analog, called a positron, has the same mass and magnitude of charge, but its charge is positive. Likewise, positive protons are paired off with negative antiprotons, and so on through all the elementary particles of matter. Each of these particles possessing mass, m, also has associated with it a minimum energy, E, given by Einstein's famous relation $E = mc^2$, where c represents the speed of light, 300,000 kilometers per second (186,000 miles per second). Any photon, to transform into a particle and its antiparticle, must be energetic enough to provide at least their combined rest-mass energies. And, of course, the distribution of photon energies is described by a temperature. So, when the universe was sufficiently hot (see Table II–3), the photons it contained could and undoubtedly did "evaporate" into particles of matter and antimatter.

The transformation was certainly not all one way. Matter and antimatter cannot coexist. When they come into contact with each other, they annihilate, disappearing in a flash of radiation characterized as a photon. In the early moments of the universe, when its density was highest, the two types of matter

TABLE II-3. *Properties of Early Universe's Particles*

Particle	Electric Charge[a]	Rest-Mass Energy	Mean Life	Threshold Temperature[b]	Age for Creation[c]
Photon	0	0	Stable	0	Any
Electron	−1	8.2×10^{-14} Joule	Stable	10^{10} K	<5 seconds
Proton	+1	1.5×10^{-10} Joule	>10^{31} years	>10^{13} K	<10^{-6} seconds
Neutron	0	1.5×10^{-10} Joule	920 seconds	>10^{13} K	<10^{-6} seconds

[a] In units of 1.6×10^{-19} Coulomb.
[b] Minimum temperature in Kelvin degrees photons must have to create the particle and its antiparticle.
[c] Age of universe since Big Bang after which photons are too cool to produce particle-antiparticle pairs.

could scarcely have avoided each other. So neither type would remain today unless one, matter or antimatter, had had a slight excess over the other. As it happened, matter had the excess; it eventually ran out of antimatter with which to annihilate. Had it not, the universe would have consisted of radiant energy only. The material structures whose shapes elicit our awe would never have come into existence. Nor, for that matter, would have we. Only the fact that the embryonic universe contained an almost insignificantly larger amount of matter than of antimatter left anything other than pure energy from which life could subsequently assemble itself.

A second cosmic circumstance affecting life's chance for ex-

FIG. II-9. Saturn. Nothing as beautiful would ever have come into being unless the early universe had been richer in matter than in antimatter. (NASA/JPL.)

isting involves the rate of expansion of the universe. Assembly of matter into living organisms would never have occurred had the universe's initial expansion rate been significantly different. Only the speed (measured in minutes) with which the cosmos passed through a thermal regime appropriate for the fusion of simple nuclei into complex ones prevented the cooking of *all* matter into forms too stable to yield energy when compressed into still heavier nuclei. The light nuclei that did survive the events of these initial moments could *later* yield energy via fusion, and that energy could drive the chemical reactions that characterize living things. In other words, the universe grew initially at a rate just sufficient to conserve some of its matter in a form useful as fuel. For that, the spectrum of life from aardvarks to zebras, azaleas to zinnias can be thankful.

The deceleration of the cosmic expansion is, as we have mentioned, determined by the mass density present. Is there significance in the fact that this parameter is close to the value separating eternal from transitory universes? Yes, for if, on the one hand, the mass density had been, say, a few hundred times the critical value necessary to bring the universe back upon itself, then the span of time granted nature to construct living systems would have shrunk to a small fraction of the time actually utilized. This might have been too brief an interval. Who is to say? But the scheme is reminiscent of an attempt to produce a human baby in one month by impregnating nine women. On the other hand, had the mass density been a small fraction of the critical value, expansion would have been so rapid that all atoms would have dispersed beyond the ranges of each other's gravitational influence. Then no aggregates larger than individual atoms could ever have arisen. The universe would have been an attractive laboratory for atomic physicists, but atomic physicists would never have appeared to work within it.

Even the amazing isotropy of the cosmic expansion worked to the advantage of the creatures who would appear. Strong anisotropies would have created pockets of exceptional density and others exceptionally sparse. The fate of the dense concentrations would have resembled a speeded-up version of a high-mass-density universe, and the rarefied regions would have resembled a low-density universe. The combination would have excluded the presence of life.

If in some make-believe world descendants could choose their ancestry, earth's biota and its hypothetical kin elsewhere would be wise to select a cosmic antiquity very much like the one they did inherit. To some, this is evidence that events in the past conspired to ensure life in the present. To others, the presence or absence of life has no effect on the quality of a universe, so a conspiracy is an unnecessary happening. Furthermore the implication that events in one epoch could have influenced those in preceding epochs turns science on its head. Nevertheless, an eternally lifeless universe poses some problems of conception. How, in the absence of anything to observe the universe's existence, could it be distinguished from something that never did exist? What is the difference between events taking place in a space-time in which nothing notes their occurrences and no events, not even any space-time, ever taking place? Perhaps some type of living intelligence is a necessity for rescuing the universe from oblivion.

Until now our discussion has considered only the gross features of the universe. An overall mean, however, is at best a skeletal description of reality. More detail is needed if we are to relate earth's living organisms to the cosmos they reside in. In the following, we emphasize a few of the highlights.

At the first level below that encompassing the whole of the universe we glimpse a spongelike structure. Strongly flattened and highly elongated assemblages of matter—sheets, pancakes, or filaments—exist as isolated structures separated by voids in space of equally huge dimension. A typical superstructure or void spans nearly one percent of the distance to the edge of the visible universe. Because of this alternately dense and sparse distribution of matter, our notion of homogeneity requires some refinement. As in a sponge, samples too small

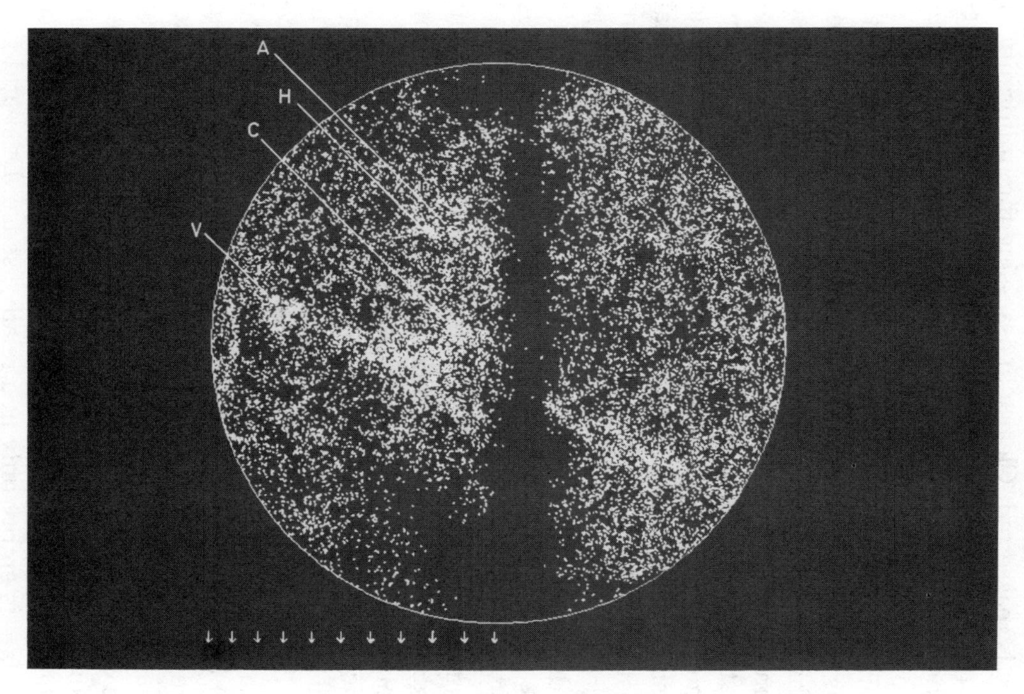

FIG. II-10. The distribution of galaxies. Antlia, Centaurus, Hydra, and Virgo clusters of galaxies indicated. Large, dark, vertical strip down center is region obscured by plane of galaxy we reside in. Some superclusters appear to cross entire sky. Dark patches, voids, also exist. (© O. Lahav and D. Lynden-Bell, I.O.A. Cambridge.)

in volume may alternate between full and empty—inhomogeneity on its grandest scale. Yet sufficiently large samples have the same relative proportions of holes and sponge, and homogeneity is reasserted. Clearly in assessing the uniformity of the universe we must choose samples whose size is a few percent of the total. When we do, the contents are the same regardless of where in the universe they are drawn from.

The contrast between the superstructures and the background is not great: their mean density is only a few times greater than that of the universe as a whole. That is because they themselves are mostly empty. Within them galaxies are concentrated into clusters, leaving relative vacuums between. Astronomers have accordingly named then superclusters. The supercluster in which we reside, parochially called the Local Supercluster, is ninety-five percent empty; ninety-eight percent of the galaxies are contained in just eleven clusters. One cluster in Virgo dominates the others. It contains one-fifth of the most luminous galaxies in the Local Supercluster, within which it is centrally located. Surrounding clusters, our own included, feel pulled gravitationally toward it. Clusters like Virgo stand out sharply above their substrate. Their densities exceed the mean of the universe by factors of thousands.

Densities are enhanced by another few thousands in the structural units—the galaxies—that populate clusters. Near the center of the centrally located Virgo Cluster lies a massive gravitational anchor called Messier 87 by optical astronomers and Virgo A by their radio brethren. Few galaxies are so large, and not all are as spherically symmetrical. Again, only a small fraction of galaxies have as active a center as M87/Virgo A. Our own galaxy, the Milky Way, is by contrast distinctly inactive.

Close examination reveals that galaxies are porous as well; stars, often grouped in clusters, dot their interiors. The space between stars is empty only in a relative sense. Dust and gas, mainly hydrogen, fill the interstellar space to a density that varies by an order of magnitude among galaxies of different types. Even though interstellar space is much more polluted

FIG. II-11. Virgo cluster of galaxies. The dominant cluster within our Local Supercluster. An example of a loose, irregular cluster containing thousands of galaxies. Their velocities spread over 650 kilometers per second, implying a mass in excess of 100,000 billion suns (10^{14} solar masses). (National Optical Astronomy Observatories.)

than intergalactic space, its emptiness cannot yet be approached by vacuums achieved in earthly laboratories.

With even the finest telescopes humans have devised, stars still appear as mere pinpoints of light.[7] Anything smaller becomes grit slipping beneath an astronomer's cosmic perception. At this level of detail, however, interest flows from the opposite direction, that of the geo-, or earth-centered, scientist. Planets, the nine members of the solar system and the suspected mul-

[7] Very special techniques have revealed surface details on a few stars.

titude associated with other stellar systems, are the next structures of significance in our descending scale. Their dimensions are mere hundredths of stellar dimensions, their masses even tinier fractions. Yet the difference between planet and star, not obvious from a casual glimpse of the sky, is fundamental: stars radiate energy generated internally, whereas planets shine mostly by virtue of the stellar light they reflect.

Zooming downward in size by a factor of a million from planetary dimensions, we enter the realm of living organisms. A good seven orders of magnitude, or powers of ten, delimit the

Fig. II-12. (a) Galaxy Messier 87 or Virgo A. A giant elliptical galaxy dominating the Virgo Cluster of Galaxies. Long exposures reveal globular clusters, the many dots on the galaxy's periphery, each containing hundreds of thousands of stars. A short exposure (lower right) reveals the peculiar jet emerging from the galaxy's center. (National Optical Astronomy Observatories.)

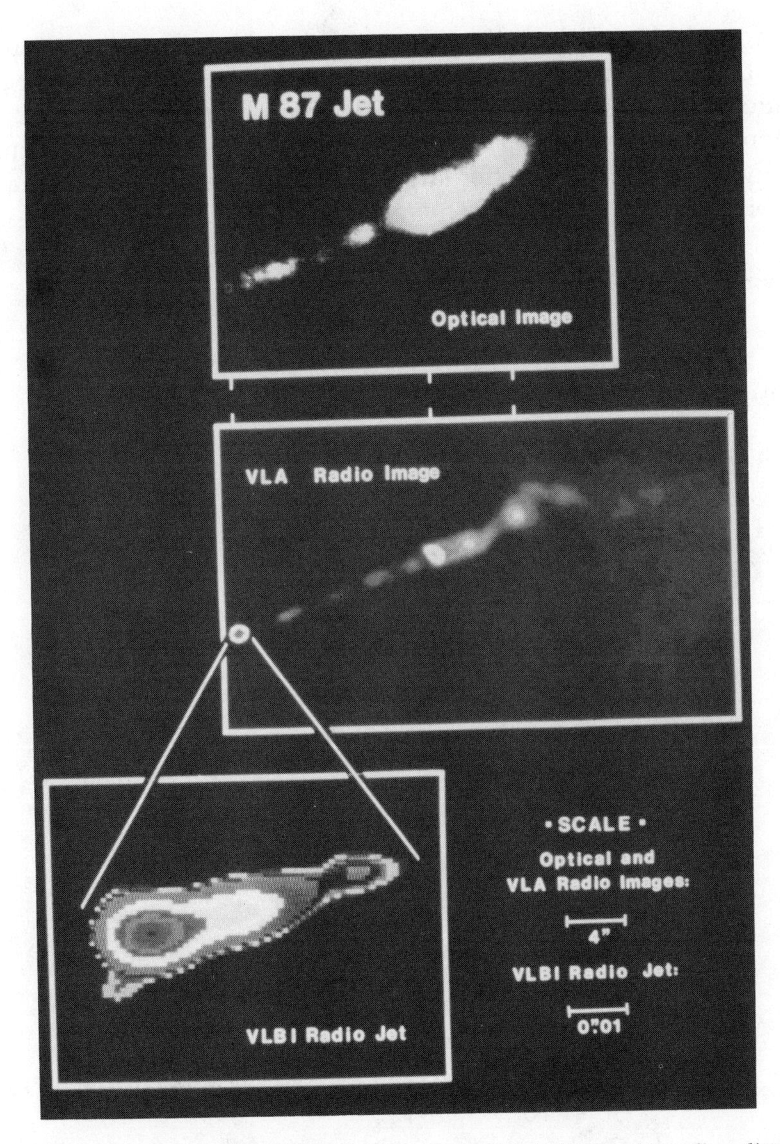

Fig. II-12. (b) The jet of Messier 87 revealed in optical light (top) and in radio radiation (center). The radio core from which it emanates is blown up at bottom. (National Radio Astronomy Observatory.)

range within which earthly organisms reside. A fly, at ten thousand times the size of a bacterium and a thousandth that of a blue whale, resides near the midrange. To place human dimensions in the spatial units of the universe, consider that the universe is the same factor larger than the orbit of Pluto as that orbit is larger than man. The factor is ten trillion (10^{13}).

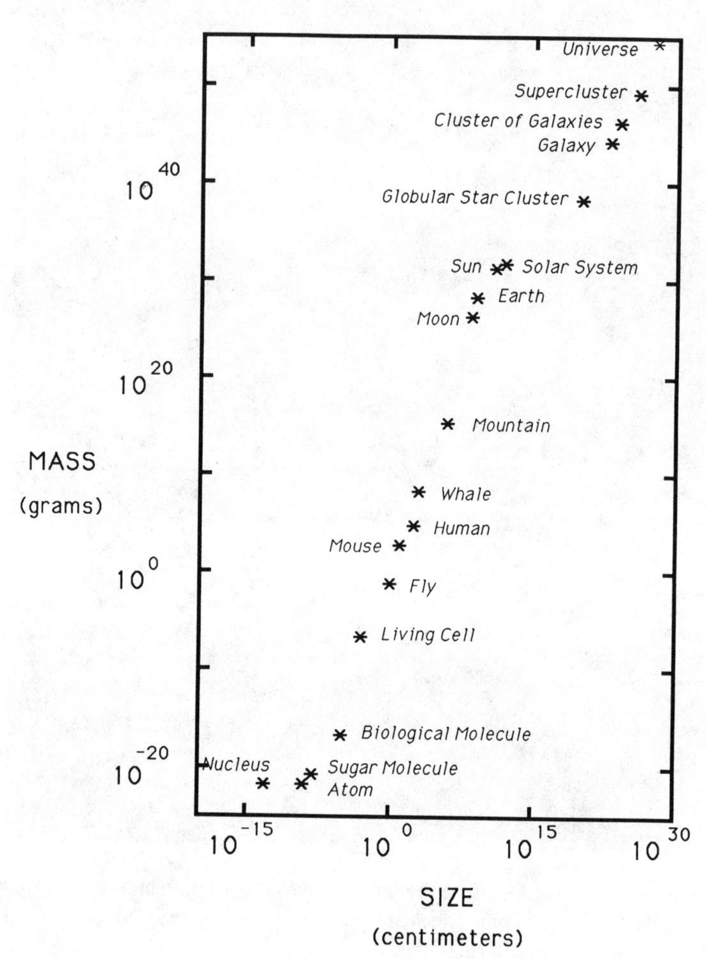

FIG. II-13. *The structures of the universe.*

We can continue peeling layers off the onion. Life is built up from molecules, whose building blocks are atoms, at whose kernels are protons and neutrons (together called nucleons) and in whose periphery are electrons. These last bits of matter are not indivisible either.

Not too far from the middle of this enormous span of sizes and masses lies the one configuration of matter that has pushed the investigation in both directions. Humans are a rough mean between the near-infinitesimal and the near-infinite.[8] Yet their significance is not merely average, for without them—or their equals or superiors in intelligence elsewhere—the whole knowledge of what is, has been, and will be, vanishes. And with the disappearance of knowledge may go the reality of the universe.

For Further Reading

Davies, Paul. 1980. *The Runaway Universe*, pp. 33–52. New York: Penguin Books.

Davies, P. C. W. 1982. *The Accidental Universe*, pp. 1–59. Cambridge, England: Cambridge University Press.

Grobstein, Clifford. 1974. *The Strategy of Life*. 2nd edition, pp. 16–33. San Francisco: W. H. Freeman and Co.

Kraus, John. 1980. *Our Cosmic Universe*. Powell, Ohio: Cygnus-Quasar Books.

McMahon, Thomas A., and Bonner, John Tyler. 1983. *On Size and Life*. New York: Scientific American Books, Inc.

Morrison, Philip; Morrison, Phylis; and The Office of Charles and Ray Eames. 1982. *Powers of Ten*. New York: Scientific American Books, Inc.

[8] For the mathematically minded, a human's size is the approximate geometric means of the sizes of a planet and an atom, that is,

$$\text{human's size} \simeq (\text{planet's size} \times \text{atom's size})^{\frac{1}{2}}.$$

His or her mass is the approximate geometric mean of the masses of a planet and a proton:

$$\text{human's mass} \simeq (\text{planet's mass} \times \text{proton's mass})^{\frac{1}{2}}.$$

Muller, Richard A. 1978 May. The cosmic background radiation and the new aether drift. *Scientific American* 238(5): 64–74.
Waldrop, M. Mitchell. 1983. A flower in Virgo. *Science* 215: 953–955.
Waldrop, M. Mitchell. 1983. The large-scale structure of the universe. *Science* 219: 1050–1052.
Weinberg, Steven. 1985. Origins. *Science* 230: 15–18.

THREE

The Times of Our Lives

It is man's potential to try to see how all things come from the old intense light and how they pause in the darkness of matter only long enough to change back into energy, to see that changelessness would be meaninglessness, to know that the only way the universe can show and prove itself is through change. His job is to do what nothing else he knows of can do: to look about and draw upon time.

William Least Heat Moon

TIME IS AN ESSENTIAL PARAMETER in any description of the universe. One reason this is so is that the universe and its contents are constantly changing. Galaxies in young clusters, for example, continually shift positions, seeking places where the gravitational energy content of the entire configuration will reach a minimum. In this quest they act like rocks on a hillside, destined eventually to arrive at the bottom of a valley, where their potential energy content is minimized. Inside individual galaxies, change is also evident. Some funnel matter toward their centers, fueling spectacular explosions seen as bursts of electromagnetic energy. Even quiescent galaxies change; they certainly age as the stars within them do so. At birth these stars incorporate matter from the interstellar medium into their dense configurations, retaining it until their death spews matter, now chemically enriched by a lifetime spent processing light elements into heavy ones, back to the spawning grounds of the next generation of stars. As a result, galaxies grow richer in heavy chemical elements as they age.

Closer to home we are also surrounded by the currents of change. Our planet is a restless one, its surface and its ocean floors recording the interplay between the buildup of freshly deposited layers and their subsequent erosion. The biota inhabiting the planet likewise define timescales that record change. Individual organisms of any species live only for an infinitesimal moment on a cosmic clock, whereas the species as a whole may persist for many millions of generations; but even the species must end, usually because its modified descendants are better matches to the environment they help create. As a distinguished paleontologist has pointed out, "all spe-

Fig. III-1. Hercules cluster of galaxies. The many galaxies in this system are still seeking a stable configuration. (National Optical Astronomy Observatories.)

cies that have ever lived are, to a first approximation, dead."[1] The same paleontologist and a colleague have even suggested that extinctions terminating many species simultaneously reoccur with a regular periodicity of approximately thirty million years. Thus another natural rhythm enters our consciousness.

All these examples of change—and only a partial list has been attempted—show that any description of the universe must include the epoch when the observations were made. As a matter of fact, citing change as a reason for adding a temporal parameter to the universe's dimensionality is circular logic: in

[1] David M. Raup, *The Nemesis Affair*, p. 47.

Fig. III-2. The galaxy NGC1275; also known as radio source Perseus A or 3C84. Compare the turbulence of this galaxy with the sedateness of NGC5364 (Figure III-5). The turbulence in NGC1275 is caused by recurrent explosions at its center. (National Optical Astronomy Observatories.)

the complete absence of change, the universe would be timeless.

Time is fundamental in another sense. The characteristics of every object in the universe are determined by the surroundings from which they are drawn, and those surroundings are modified by the emergence of newly formed objects within them. In other words, the environment evolves because its inhabitants do, and the inhabitants evolve because their environment does. Examples of objects exquisitely tailored to their environment abound among earth's biota. To cite just one, consider the exotic organisms thriving at rifts in the sea floor. These use the noxious gas hydrogen sulfide, which is vented from

Fig. III-3. Earth's dynamism is exposed at Titus Canyon, Death Valley National Monument.

the rifts, as the energy source enabling them to transform carbon dioxide into the substance of their bodies. By way of contrast, land plants use sunlight as the energy source to drive their utilization of carbon dioxide. Neither lifeform could survive in the other's environment, but each can take advantage of the resources available in its own environment. Examples of conformity to a setting are not confined to living organisms, either. The inner four planets of the solar system are smaller, denser, rockier, and have fewer light elements in their atmospheres than the outer five,[2] all because of the different thermal and gravitational regimes in which they formed. Likewise, the centers of galaxies, where relatively high density encourages rapid star formation, have higher proportions of old stars than do the peripheries, where some potential stellar matter has been preserved in a dispersed form. Even within superclusters,

[2] Pluto, about which little is known, may be an exception.

a galaxy's location and the neighbors with which it interacts determine its type. Most galaxies in the deepest gravitational wells centered on the densest clusters are noticeably bulged. An example is Messier 87 (Figure II-12). It and other elliptical galaxies contain predominately old stars and little gas or dust between. Farther from the foci of clusters, the majority of galaxies are gas-rich, populated with young stars, and highly flattened. Again, the causes of the differences are environmental.

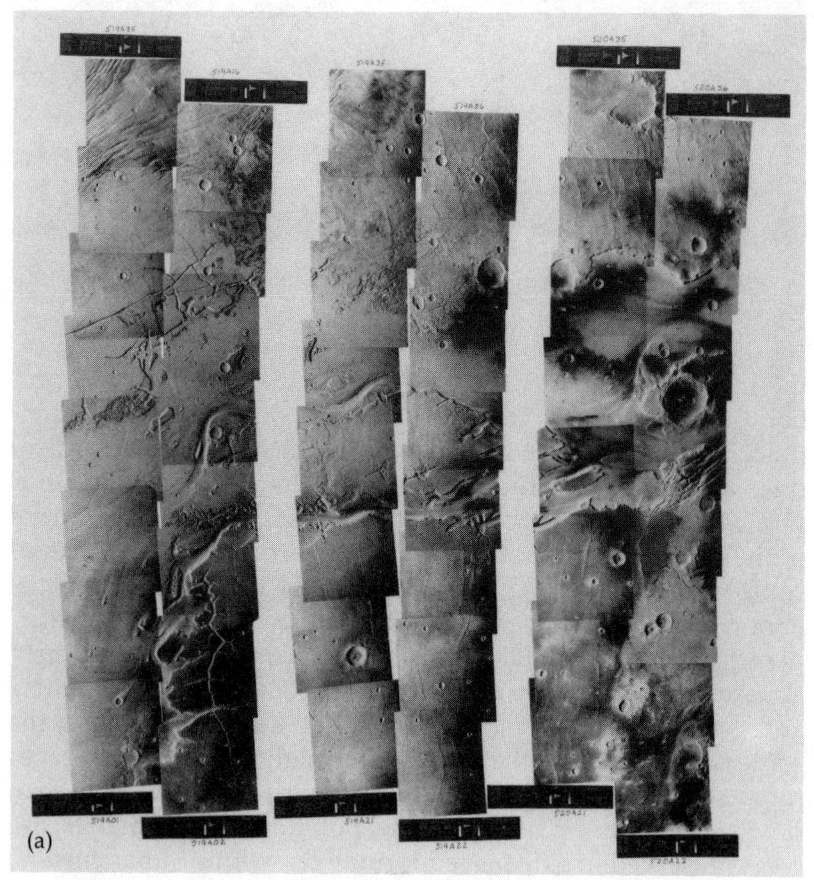

FIG. III-4. Rocky surface of an inner planet, Mars (a), contrasted with gaseous atmosphere of the outer planet, Jupiter (b). Jupiter has no solid surface. (NASA/JPL.)

Fig. III-4(b).

Finally, we restate the reason time is fundamental: since everything is the product of its environment and since the environment is always changing, *what an object is depends upon when it comes into being*.

What portion of the surroundings of a newly forming object is relevant to defining its structure and composition is determined by how far and at what rate information travels to its site of formation. How much time has been available for communication? Which objects have had the opportunity to exchange information? Size clearly helps determine the relevant timescale. For just one instance, superclusters are so huge that the time required for a galaxy, traveling at a typical observed speed, to cross from one end to the other exceeds the age of the universe. Consequently superclusters have not been thor-

Fig. III-5. NGC5364. In a cluster of galaxies, spirals of this type are more common around the periphery than near the center. (National Optical Astronomy Observatories.)

oughly mixed. They have not assumed their final configurations. The supercluster environment continues to work on its membership, changing and molding its collective distribution. One senses time marching on—at a cosmic pace in this instance.

Of course, information floods a small region faster than a large one. As a result, times characteristic of different physical systems are gauged in different ways. In the preceding chapter we saw that the universe possesses structure in a hierarchy of sizes and that each layer of the hierarchy is made up of objects on the layer immediately below: atoms constitute stars, which compose galaxies, the building blocks of clusters, which are themselves the units of superclusters. Since time and size are interdependent, one expects durations of time also to occupy a hierarchy. And indeed they do.

Since this is a book about the *living* universe, we shall enter the temporal hierarchy at the level appropriate for biological time scales. Here we find a succession of processes, each building upon a huge number of processes quicker than itself. At the smallest, hence quickest, relevant level are *molecular processes*. These involve chemical reactions, which can occur as quickly as a hundred-thousandth (10^{-5}) of a second or as slowly as a second. Billions of such molecular interactions constitute a *physiological process*, say the transmission of a nerve impulse or the contraction or extension of a muscle. Relevant times for physiological processes range from a hundredth (10^{-2}) of a second to an hour. Of course, millions of muscular flexings guided by a nervous system are required to perfect a skill. So *developmental processes* may require minutes or decades, depending upon the size of the organism and the difficulty of the skill being mastered.

The lifetime of an individual organism includes any number of developmental processes. Time intervals are fundamental in biological processes, for an organism *is* a complete life cycle. Reproduction does not duplicate adults; it produces the single cell from which full-sized offspring are constructed via cell duplication and differentiation. Until the new organism attains the maturity of its parents at the time of their union, their full replacement has not been accomplished. During the whole of the time that an offspring is maturing to adulthood, its genes are shuffling and reshuffling. As they do, the diversity of options for possible survival in a competitive environment is enriched. It is clear, then, that the rate of evolutionary change depends upon the generation time, or the time for *reproductive processes*. Naturally it takes more time to build large structures than small ones, reproductive processes therefore varying from tens of seconds to hundreds of years. A lifetime of half a minute may, to us, seem devoid of opportunity, but only because of

Fig. III-6. Developmental processes. The learned skills of these dolphins required many months of training.

our strong anthropomorphism. In reality—that is, measured by the beats of their own internal clocks—most organisms have comparable lifetimes. An elephant's weight exceeds a pika's by a few hundred thousand times, but the elephant needs to generate much less heat *per unit of weight* than does the pika to maintain a constant internal temperature; so the elephant's metabolic rate can be much lower than the pika's. If we measure metabolism by breathing rate or heartbeats, we find that all mammals' lifetimes consist of about two hundred million breaths and eight hundred million heartbeats.[3] These can either

[3] For the reader, nervously counting his pulse, humans are exceptions. We live roughly three times longer than mammals of our size "should." We are granted three times as many breaths as a comparably sized bear.

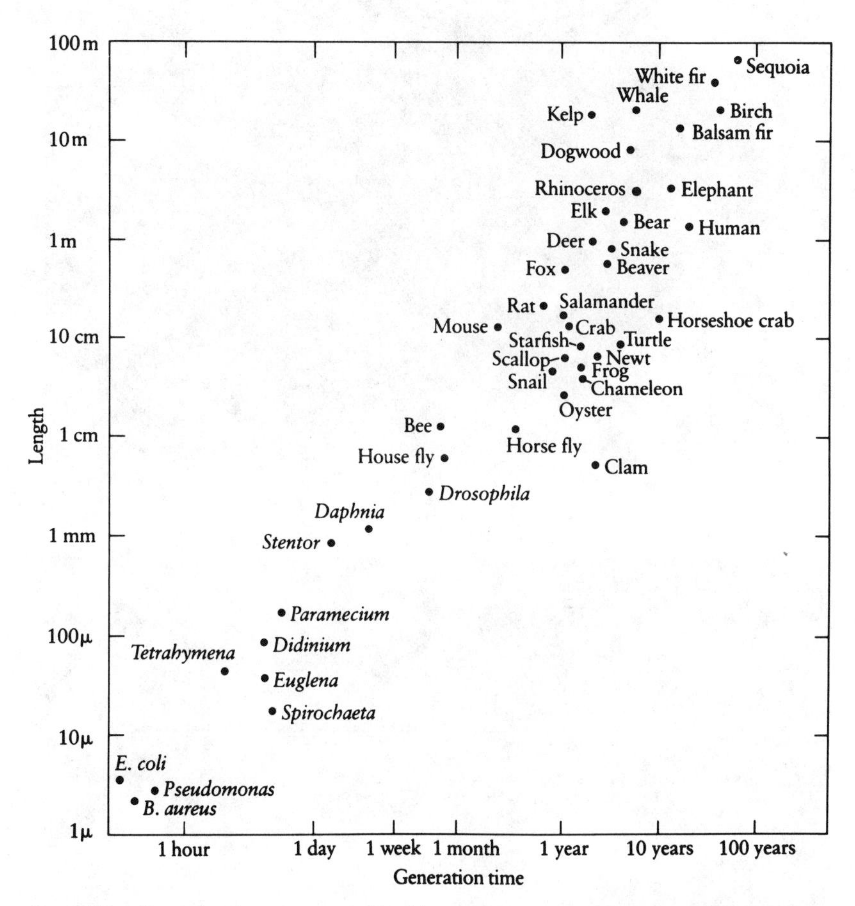

Fig. III-7. Reproductive processes. The bigger an organism at the time it reproduces, the longer its generation time. (From John Tyler Bonner's Size and Cycle, *Copyright © 1965 by Princeton University Press.)*

be expended frenetically, say fifty heartbeats per second for the hyperactive pika, or leisurely, a single heartbeat per second for the ponderous elephant. Neither is to be envied or pitied, since each lives for the same number of beats of its own heart. The elephant lives fifty times longer than the pika, but the sum total of its activity is no greater.

The notion that different organisms experience the passage

Fig. III-8. Life's pace. Despite the enormous size difference, both the elephant and the pika will enjoy about the same total numbers of breaths and of heartbeats in their lifetimes. The elephant, though, will live fifty times longer by expending his at a slower rate.

of time at different rates delivers a poignant message. A mayfly lives a mere day, which it may experience as fully as we do our lifetimes of nearly a century. But the mayfly's interpretation of events lasting several days, must differ sharply from our own. Only an exceptionally wise mayfly could know that a butterfly was previously a pupa and before that a larva, for instance. It should be apparent to us humans, therefore, that some cosmic processes, because of their extremely tardy paces, will yield to our comprehension only with the greatest difficulty. Even our existence as a species is so brief compared to all the aeons of history that many cosmic secrets must elude our recognition. Even more to the point is the fact that we may be engineering massive changes on our own planet faster than its geologic clock can respond to them.

At the next temporal hierarchy above generation times are *historical processes*. An historical process for a particular species may last several generations and involve, in the end, millions of individuals. An example underway at present, although still unfinished, is the spread of the Dutch elm disease across North America. Its completion will take many decades, perhaps more than a century. Simultaneous with it, other equally large-scale historical processes are occurring: condors are disappearing for all time from California, as are ferrets from the Great Plains. The outcome of these and other simultaneous historical processes is an *evolutionary process*, the introduction of a new genetic mixture. Relevant timescales for the introduction or elimination of species can be tens of thousands to millions of years (10^{11} to 10^{13} seconds). Major extinctions of many species followed by explosive radiations of opportunistic ones may even recur periodically at thirty-million-year intervals. Of course, the more time that is available, the greater the numbers of individuals, species, and total biomass that can get involved. Whole ecosystems can therefore change over periods some tens to hundreds of times longer than evolutionary processes require.

At the grandest biological scale consider, as a unit, all the biota inhabiting earth's global environment. Together they constitute a single superorganism, a coordinated whole, a point we shall develop in some detail in Chapter VI. The superorganism, like each individual organism, began its existence as a single cell nearly four billion (4×10^9) years ago and has been multiplying, diversifying, and growing ever since. So recently that the epoch can be referred to as Now, it brought forth a creature endowed with the ability to carry the superorganism's lifestream elsewhere, i.e., the human being, who is able to travel and reside elsewhere in space. Furthermore, these same human members of the superorganism may perhaps be able to deduce the ultimate fate of the universe and crudely to outline steps necessary to perpetuate the lifestream. These accomplishments indicate life has arrived at a significant threshold. The time required, some few billions of years (10^{17} seconds), consequently can be used as the timescale for superorganisms, or global biological communities, to achieve maturity sufficient to comprehend themselves and to carry their presence beyond their place of origin. Earth's life is fortunate to have persisted this long, because it now contains the wisdom to last much longer.

Time also manifests itself in the universe through the motions of material bodies and the forces that govern these motions. Starting at the top of the structural hierarchy, we can ask how much time is necessary for the universe as a whole to move into a new configuration. Since gravity is the predominant force shaping the universe, it is appropriate to seek a timescale for a gravitating system. It is found in the so-called *free-fall time*, the time for a system of distributed masses to implode upon itself. The free-fall time depends upon two factors: the amount of matter or mass, and the volume of space over which it is spread. These two factors can be combined into a single parameter, density, whose units are mass per unit volume (for example, grams per cubic centimeter). The strength of the attraction between two objects varies as the product of their masses; this is where the amount of matter gets involved. The

Fig. III-9. A human in orbit about the planet of his birth. A preparatory step to carrying the lifestream elsewhere? (NASA.)

attraction also diminishes inversely as the square of their separation,[4] which explains how volume is involved. Clearly, massive objects close together collapse to a single lump faster than less dense configurations. When the calculations are quantified, one finds the free-fall time of the universe to be approximately

[4] Doubling the separation reduces the attractive force to one-fourth its former strength; tripling it, to one-ninth; and so on.

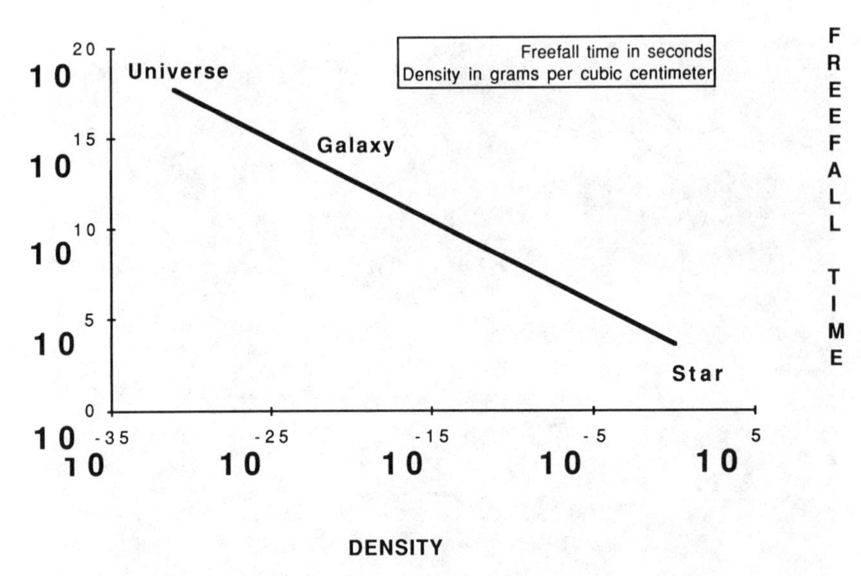

Fig. III-10. Freefall time of distributions of matter as a function of their density.

100 billion (10^{11}) years (10^{18} seconds). This duration exceeds the age of the universe, but only by a factor of ten. Had the same matter occupied space less extravagantly—recall that, on average, every cubic meter or dump-truck load of space contains only a single hydrogen atom—the universe would already have completed its free-fall, and the structures (galaxies, stars, and planets) we now recognize as necessary progenitors would have been smashed together and incinerated in the process. The sheer immensity of the universe has been a major factor in permitting the emergence of life within it. Space has been a nursery only because there is so much of it.

Size is not the only benefactor contributing to our existence. Gravity has pulled together ensembles called galaxies whose densities exceed the universe's average by a million (10^6) times and whose free-fall times are consequently reduced by a factor of a thousand (10^3): 100 billion years for the free collapse of the universe shrink to only 100 million for the collapse of a galaxy.

But life has existed on our planet some tens of times longer than this, so other factors must postpone the galaxies' demise. Prominent among these is *rotation*. An object—star, planet, satellite, whatever—can fall toward the center of a system only until the centrifugal force it receives from its rotational motion balances the centripetal pull of the system's gravitation. It can remain in orbit at that null position for a very long time (although, because of tidal friction and chance encounters with other objects, not forever). For example, the sun moves around the center of the Milky Way galaxy with a period of about 250 million years (2.5×10^8 years, which is equal to 10^{16} seconds), and is consequently able to maintain its distance from the center at ten billion trillion (10^{22}) centimeters. Just as we use the earth's revolution around the sun as a measure of time appropriate for planetary systems (the year, or thirty million seconds), we can use the sun's orbital motion to define a timescale for galactic systems. The fact that "galactic years" measured in this way are much longer than free-fall or collapse times for those same galaxies is a fortunate circumstance. Rotation is another protector for whatever life dwells in galaxies, for without it there would be insufficient time for life to establish itself.

Nature erects yet a third barrier against the inward pull of gravity, the *thermonuclear barrier*. Its necessity is evident: the free-fall time for matter in a star is only an hour or so, and none rotate fast enough to generate an equivalent centrifugal tendency. Heat, though, boils from the stars' bowels, and its flux presses the overlying layers outward. The heat is a byproduct of the fusion of quartets of hydrogen nuclei into single helium nuclei. The force binding the four particles in the latter nucleus is so strong that they gain more energy than they possessed individually by falling into its embrace. That energy gain, of course, cannot be contained within the fused nucleus, but is instead released and used to oppose the star's compression. Life profits directly from the thermonuclear barrier. Not only does it prevent matter from collapsing into vanishingly small points, but also it does so with just the proper opposition to

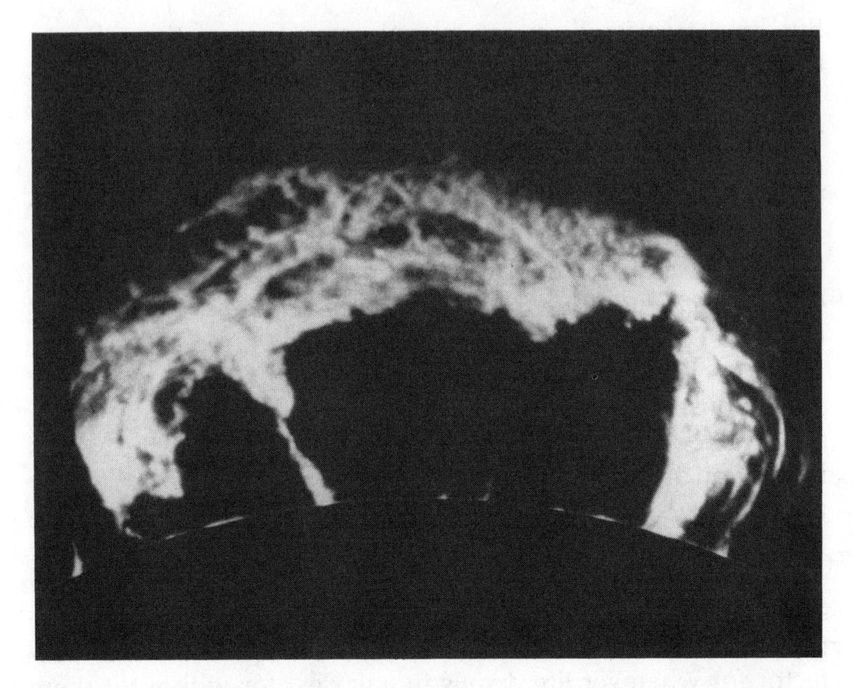

Fig. III-11. Prominence erupting from surface of sun, a dramatic manifestation of the release of energy generated at its center. (National Optical Astronomy Observatories.)

preserve stable flows of energy over time spans long enough to permit living systems to parasitize the flows. The sun, for instance, has faithfully delivered energy from the inferno of its innards to the coldness of its surrounding space for five billion

Table III-1. Energetics of Thermonuclear Fusion

Mass, 4 unbound hydrogen atoms	4.03188 atomic mass units
Mass, 1 helium atom	4.00260 atomic mass units
Mass Decrement	0.02928 atomic mass units
Energy Equivalent[a]	0.00004 ergs

[a] The energy of a fly landing on the page you are reading is approximately 1 erg.

(5×10^9) years, and will do so for an equivalent time into the future. Its expected lifetime (10^{10} years, or 10^{17} seconds) is comparable to the present age of the universe. Had it been significantly less, matter on earth might not have had sufficient time to arrange itself into forms capable of utilizing solar energy for the life-giving tasks of duplication, growth, and metabolism.

Why is the consumption of hydrogen so stately? When humans first engineered fusion devices, they constructed hydrogen bombs that consumed fuel almost instantaneously. What prevents the sun from exploding like a bomb, thereby robbing life of the incubation time it requires? The answer lies in the magnitudes of the two kinds of forces, called strong and weak, which act between the particles in the nuclei of atoms. The strong nuclear force, whose effects were mentioned earlier, is the one that binds nucleons (protons and neutrons) together. The weak force governs changes in the particles' identities rather than in their motions. It regulates radioactive decay. In particular, it governs the rate at which neutrons decay into protons.[5]

Our feel for these two types of forces is not intuitive, since neither can be experienced directly. Only the advent of particle accelerators (see Figure II-8) permitted their study, because only with such accelerators could particle interactions over extremly short distances (less than 10^{-13} cm) be studied. The need for both forces, however, can be inferred. A strong force, one can see, has to exist to keep nuclei from flying apart. Since roughly half of a nucleus's particles are positively charged (the remainder being neutral), and since positive charges repel each other increasingly as they are brought closer together, some "glue" must exist to overcome this repulsion. As for the weak force, it too was called for when radioactivity was discovered. Without it, scientists were at a loss to explain how a uranium nucleus could emit two protons and two neutrons, and thence be trans-

[5] Half of a collection of free neutrons decays into equal numbers of protons and electrons in 750 seconds.

Table III-2. The Fundamental Forces of Nature

Force	Application	Relative Strength	Range (cm)
Strong	Binds atomic nuclei	1	10^{-13}
Electromagnetic	Binds neutral atoms	1/137	Infinite
Weak	Governs particle decay	10^{-5}	10^{-15}
Gravity	Curves space time	6×10^{-39}	Infinite

formed into a thorium nucleus. The term weak is relative: the strength of the weak force is intermediate between those of gravity and electromagnetism. But a key concomitant of that weakness is that weak-force interactions between particles occur slowly. Uranium nuclei are not decaying instantaneously, nor neutrons either, for that matter.

Consider the consequences of altering the strengths of the two types of force. Imagine, first, that the strong nuclear force were stronger—by as little as a few percent! Then two protons could fuse together into a nucleus of helium-two (^2He). Since no transmutations of matter would be necessary, only the strong force would be involved. The reaction could therefore proceed quickly—so quickly that all the hydrogen in the sun would long ago have been consumed. Life on a planet near such a short-lived star would either never have arisen or would already have been extinguished. But this horror has been prevented because the strong force cannot quite bind two protons against the electromagnetic repulsion they exert upon each other. The solar conversion of hydrogen to helium must therefore occur in a more roundabout way. In its very first step, the two protons must create a nucleus of heavy hydrogen (^2H) called deuterium. One of the nucleons in the fused nucleus must have no electric charge, so that neither of the pair feels repulsed by the other. The weak force comes into play, governing the necessary conversion of protons into neutrons. Being weak, the force has difficulty engaging particles. In other words, reactions using it are slow—a billion billion (10^{18}) times slower than a strong nuclear reaction among particles at the same density and temperature. Some hydrogen fusion in stars has been postponed until life can make appropriate use of it precisely because strong interactions are not too strong and weak interactions are benevolently slow.

Tampering with the strength of the weak force would also introduce serious consequences. Recall from the discussion in the last chapter that the universe spent a few minutes shortly after its creation in a temperature and density regime as suitable for the synthesis of helium from hydrogen as is found today only in stellar cores. When the universe's temperature and density were appropriate for helium fusion, about thirteen percent of all nucleons remained in the form of neutrons. The proton pool, which started equal to the neutron pool, had been augmented by the decay—via the weak interaction—of nearly three-fourths of the initial neutrons. As a result, the thirteen percent of nuclear matter remaining as neutrons united with another thirteen percent in the guise of protons to lock twenty-six percent of the nuclear material into stable helium nuclei, each consisting of two neutrons and two protons. The remaining seventy-four percent was preserved as hydrogen nuclei. These could—much later to be sure—be utilized as fusion fuel inside stars. And the energy-radiating stars could provide suitable environments for the creation and propagation of life. If the weak interaction had permitted less rapid conversion of neutrons into protons, then the two types of nucleons would have been more nearly equal when the "window of opportunity" for fusion presented itself. Suppose conditions appropriate for nucleosynthesis had prevailed when the split between neutrons and protons had been, say, forty-five to fifty-five. Then fully ninety percent of the energy resources sustaining our existence would have been consumed too early for

TABLE III-3. *Characteristic Times*

System	Description of Interval Characterized	Time (seconds)
Universe	Free-fall time of all matter	10^{18}
	Age	$\lesssim 10^{18}$
Supercluster	Time for galaxy moving at average speed to cross	$\gtrsim 10^{18}$
Milky Way Galaxy	Orbital period of sun	10^{16}
Sun	Lifetime	10^{17}
Solar System	Age	10^{17}
	Orbital period of earth	10^{7}
Superorganism	Maturation time	10^{17}
Ecosystem	Time to become climax community	10^{12}–10^{15}
Species	Evolutionary process, Typical interval before extinction	10^{11}–10^{13}
Helium Nucleus	Time to fuse from hydrogen nuclei	10^{10}
Organism	Generation time	10–10^{9}
Organ	Physiological process	10^{-2}–10^{4}
Neutron	Halflife decay time into proton	10^{3}
Molecule	Chemical reaction	10^{-5}–10^{0}
Atom	Electron orbital period	10^{-16}
Nucleus	Light travel time across system	10^{-24}

us to take advantage of them. The weak force served us well. Had its strength been otherwise, neither we, nor any other life, might ever have come into existence.

Tiny changes can obviously have major consequences. The pulse and throb of living organisms exist because rhythms and cycles on vastly different timescales had values of precisely the proper magnitude. Nothing quite so vividly illustrates life's interdependence with the full dimensions of space-time. We seem products of a history determined to include us. Let us also be guardians of a future permitting our further development.

For Further Reading

Bonner, John Tyler. 1980. *The Evolution of Culture in Animals,* pp. 14–18, 57–59. Princeton, N.J.: Princeton University Press.

Davies, P. C. W. 1982. *The Accidental Universe,* pp. 9–21, 40–59. Cambridge, England: Cambridge University Press.

Dyson, Freeman J. 1971 September. Energy in the universe. *Scientific American* 224(3): 50–59.

Gould, Stephen Jay. 1977 Aug.-Sept. Our allotted lifetimes. *Natural History* 86(7): 34–41.

McMahon, Thomas A., and Bonner, John Tyler. 1983. *On Size and Life.* New York: Scientific American Library.

Oort, J. H. 1982. The nature of the largest structures in the universe. In *Astrophysical Cosmology,* ed. H. A. Brück, G. V. Coyne, and M. S. Longair, pp. 145–162. Vatican: Pontificiae Academiae Scientiarum.

Raup, David M. 1986. *The Nemesis Affair.* New York: W. W. Norton and Company, Inc.

Trefil, James S. 1983. *The Moment of Creation,* pp. 70–86. New York: Macmillan Publishing Company.

FOUR

Order from Chaos

Economists are fond of saying that "there is no free lunch," everything must be paid its value so that price and value always balance out. The entropy law teaches us that mankind lives under a harsher commandment: in entropy terms, the cost of a lunch is greater than its price.

Nicholas Georgescu-Roegen

IN DISCUSSING THE UNIVERSE, we have ordered events in a sequence establishing a direction to time. Earlier can be distinguished from later. This indicates that a structure has been imposed upon cosmic history. Why should this be so? Why do effects follow causes? Why do all events not occur simultaneously, or in a randomly sequenced order? The answers lie in various disequilibria, or gradients, that exist in the universe. Hot stars pour energy into cold space. Gas pressure disperses some assemblages of particles, while gravity pulls others more tightly together. Tidal friction slows the rotation of a satellite until one side becomes locked into position facing its central planet. Galaxies tumbling within clusters also feel tidal forces, which convert their motions from arbitrary to systematic. Closer to home, rocks fall down mountainsides, cold water settles to the bottoms of lakes, air moves from high to low pressure regions.

All these and other gradients establish slopes along which events can slide from one state to another in a "natural" direction. Experience teaches which direction is natural and consequently permits the various states to be ordered in time. For example, one can determine that the Appalachian Mountains are older than the Rockies by comparing their heights and, in particular, the steepness of their profiles. This is because no one has seen rocks spring unaided from low elevations to mountain summits, whereas the reverse is commonly observed.

The existence of gradients in the universe indicates that the

FIG. IV-1. Grand Teton (top) vs. Appalachian Mountains (bottom). The Tetons are sharper and taller, therefore younger.

matter and energy it contains are not distributed in a random, haphazard way. Space-time is not filled uniformly with matter ground to an irreducible rubble. Rather, organized structures exist on a variety of scales. Yet we believe the universe started as an intensely hot, thoroughly intermixed medium. Indeed, our farthest probe into antiquity, the three-degree blackbody background radiation discussed in the second chapter, is evidence for an era in which the constituents of the universe comingled in thermodynamic equilibrium. How, then, have disequilibria arisen, when all our experience suggests that natural events proceed *toward* the stability that total equilibrium represents? How has order arisen from primordial chaos?

No symbols of order seem, superficially at any rate, as unlikely as living organisms. The chemical constituents of the human body, for example, are known in detail. Yet, given a pile of the necessary chemicals, no one knows how to assemble them into a living being. Ignorance is especially profound about how all the contents are integrated with each other and how, throughout the life cycle, different necessary actions are taken at the proper times. The entire complicated assemblage seems totally improbable. Contrast this impossibility of constructing a fully alive unit by stirring together its constituent parts with the ease with which such a unit is rendered dead (by stepping off a cliff, for example). Since nature normally runs in the direction of greatest ease—from improbable to probable, from complicated to simple, from ordered to random, or, in this case, from life to death—how did life establish a foothold on (at least) Planet Earth and subsequently blossom in richness, diversity, and complexity?

Expansion of the universe, particularly the rapid expansion of its earliest moments, explains how order was able, eventually and perhaps temporarily, to emerge from chaos. In effect, a system of conservation was established so that some energy was reserved in a form that could later be used to do the work necessary to impose order. Before examining this mechanism in detail, we must consider the fundamental laws, in effect throughout the universe, governing energy and its use.

The first law of thermodynamics is a familiar one: *the energy of the universe is a constant*. Energy can be neither created nor destroyed. It can, of course, change manifestations. Particles and antiparticles can annihilate each other, leaving flashes of heat and light whose energy equals that locked in the masses of the vanished grains of matter; or pairs of grains can spontaneously appear from sufficiently intense concentrations of radiant energy (see Figure I-10). Likewise, heat energy can be turned into motions of mechanical devices—witness the steam engine, while the friction inherent to mechanical devices converts some of their kinetic energy into heat. Other examples abound of this no-win, or zero-sum, situation.

Were this the only law governing energy in the universe, the processes within would be eternal. As the operation of a given physical system exploited one form of energy and discarded another, the supply from the discard pile could be continuously reconverted back into the one sustaining the operation. In this world of perpetual motion, the sense of past and future would be erased. Would the energy be flowing from an input source to a discard pool and operating a machine in a "forward" direction, or would the energy flow be reversed and the machine operating backward?

In the real world, the forward-reverse ambiguity is not present because different forms of energy have different qualities or merits. Some are more able to perform work than others. Water behind a high reservoir possesses a potential gravitational energy. Letting it fall through a spillway converts its stored energy into kinetic energy, an energy of motion, which can do work, e.g., spin a turbine. When the water strikes rocks at the base of its descent, its potential for doing work is completely exhausted. The energy of its bulk motion has been dissipated into the heat it imparts to the rocks. An orderly flow

Fig. IV-2. Cass, West Virginia, Scenic Railroad. Heat is converted into mechanical motion. Energy can change forms, but the total cannot be changed.

of water downhill has yielded to a disorderly splash at the bottom. The process is *irreversible*. Heat is not spontaneously extracted from rocks and used to drive water uphill. Water can, of course, be transported to the top, but only by expending considerable energy, the utilization of which again results in a greater transfer of energy from a useful, or available, form to a useless, unavailable one. This very irreversibility assigns a direction to time. The forward and reverse directions of a movie of a waterfall are distinguishable immediately.

Fig. IV-3. Bonneville Dam on Columbia River. Water behind the dam possesses potential energy. Water below the dam has irreversibly exhausted its potential.

Nature finds it easy to create anarchy, but difficult to impose order. This much is clear from the fact that random, disordered states vastly outnumber ordered ones. The letters on this page are arranged in an ordered fashion, one laden with information (or so the author hopes). The probability that this same configuration of letters, or any other meaningful combination, could be achieved by throwing Scrabble pieces is effectively zero. There are simply far too many combinations of nonsense compared with the few meaningful ones. When we say that natural processes proceed toward disorder, we are only stating that they seek configurations made more probable by their overwhelming numerical advantage.

The concept of *information* appears in the foregoing example. Certain arrangements of letters convey more meaningful information than arbitrarily configured ones. Ordered systems in general have a high information content. Their description requires specification of many parameters. Disordered systems, in contrast, contain little information. They are easy to describe. The difference is illustrated by comparing a message written by a skywriter using water vapor with the chaotic appearance of a cloud. The former system is ordered, full of information, and unlikely to arise spontaneously. The latter is disordered, devoid of interpretable information, and quite likely to arise spontaneously. Again, to emphasize the irreversibility of physical processes, note that the carefully crafted message will quickly degenerate into an amorphous cloud, whereas the written word never spontaneously emerges from the prevailing overcast. Any spontaneous process acting on an isolated system likewise results in a loss of information.

The equilibrium state is *the* most probable one. It requires the minimum quantity of information for description and has the greatest disorder. It represents the ultimate configuration of all physical systems left to their own devices. As long as disequilibrium exists, there can be change. Once equilibrium is achieved, change ceases. For example, heat can be a useful form of energy when it is arranged in an ordered, non-uniform way—perhaps a glowing ember radiating into a cool room. When the ember has cooled to room temperature and thermodynamic equilibrium is achieved, however, no further useful work can be extracted because there is no net flow of energy.

Entropy is a quantity physicists have defined to represent the opposite of information. High entropy implies little information content. It also connotes randomness and disorder. High entropic systems are probable ones. Low entropy is assigned to improbable, highly ordered systems crammed with information. The natural tendency for isolated physical systems is embodied in the second law of thermodynamics: *the entropy of the universe always increases*. Whereas the first law forbids winning (that is, acquiring energy from nothing), the second forbids even breaking even. When engineering any change, one has decreased the amount of energy available to perform work in

Fig. IV-4. Entropy increasing. A chair in the abandoned gold-mining town of Bodie, California, is tending toward total disorder.

the future. Note that we shall always know in which time direction the future lies—the one characterized by greater disorder.

Armed with these two grand principles of physics, we must now address how the universe could have evolved from the chaos of thermodynamic equilibrium—a system so limited in information that its description required a single parameter, the temperature—to the ordered system of organized structures whose description occupied most of the preceding chapters. The universe has behaved benignly to conserve some energy in a form retaining the potential for imposing order. It did not degrade immediately to the final equilibrium state from which no further useful enterprise could be extracted. Instead its rapid, initial expansion permitted postponement of the inevitable. Expansion produced cooling.[1] As the universe cooled, it underwent occasional changes in its state crudely analogous to the "freezing out" of, first, liquid, then solid, water from an initial vapor. Water in the state of vapor can spread evenly throughout the volume it occupies, but cooling it into liquid drops introduces a grainy structure, which is exaggerated by further cooling into ice crystals.

If the cooling occurred rapidly enough—and in the case of the universe the temperature fell from some hundred million trillion trillion (10^{32}) Kelvin degrees immediately after the Big Bang to ten billion (10^{10}) after only a few seconds (see Figure I-9)—matter and energy, initially in perfect equilibrium, never could re-achieve equilibria at successively cooler temperatures. They grew farther and farther apart, establishing exactly those disequilibria from which order could spring. Just as an example, when the universe was younger than one second, photons were so hot they had more than enough energy to supply the summed masses of resting electrons and anti-electrons (called positrons). This occurred constantly, while just as frequently electrons and positrons were annihilating to produce replacement photons. After the first second, however, photons were too cool to possess the energy equivalent to the mass of an electron-positron pair. Only annihilation could continue, and it ceased when the supply of positrons, fortunately slightly less numerous than that of electrons, was exhausted. Interchangeability of matter and energy then ceased; the ratio of electrons to photons was established for all subsequent time; a disequilibrium between matter and its energy bath had arisen. While it seems as if order had been imposed upon chaos, and hence entropy had decreased, this was not so. The decrease in entropy occurring when electrons "froze out" or materialized from their substrate was more than compensated by an entropy increase woven into the fabric of space-time itself. Expansion

[1] The inverse of this process is familiar to everyone who has ever handpumped a bicycle tire; the pump itself grows warm, heated by the air compressed inside it.

of this fabric permitted the total quantity of entropy in the universe to increase. There was always more space requiring new information for its description. The inability to provide information as fast as expansion drove the need for more of it represented an increasing ignorance. The news about some particular portion of the universe could not be communicated to all other portions before changes in the original portion made the news obsolete. This widening gap between what one *could* know and what one actually *did* know was equivalent to an increase of entropy.

Energy-matter, in existence for scarcely a second, was still a very long way from its ultimate destiny, its least available form. The next transformations en route to this goal involved nuclei. At about the same time that electrons materialized from their substrate, temperatures became appropriate for commencement of fusion of light nuclei into heavier ones. Before that time, temperatures (and densities) were so high that any compound nuclei were torn asunder by the violence of the frequent collisions they experienced. And after only a brief "window of opportunity," fusion again could not take place. To see why requires a side excursion into nuclear physics.

Among nuclei, iron is the most stable, the one in which nucleons are most tightly bound. In analogy to gravity, the iron nucleus corresponds to the lowest level to which any object can fall, the energy level devoid of potential for extracting further useful work. Iron is therefore the nucleus of highest entropy. Why, then, did not all nuclear matter sink into this most degraded form? Expansion again provides the answer. For after slightly more than three minutes, temperatures had fallen below a few billion Kelvin degrees. Nucleons consequently were no longer propelled with sufficient vigor to approach within the extremely short range of their strong forces; and slightly over three minutes was not long enough to overcome the delay introduced into progressive nucleosynthesis by the "deuterium bottleneck" occasioned by the weak nuclear force (discussed in the preceding chapter). So fusion was quenched

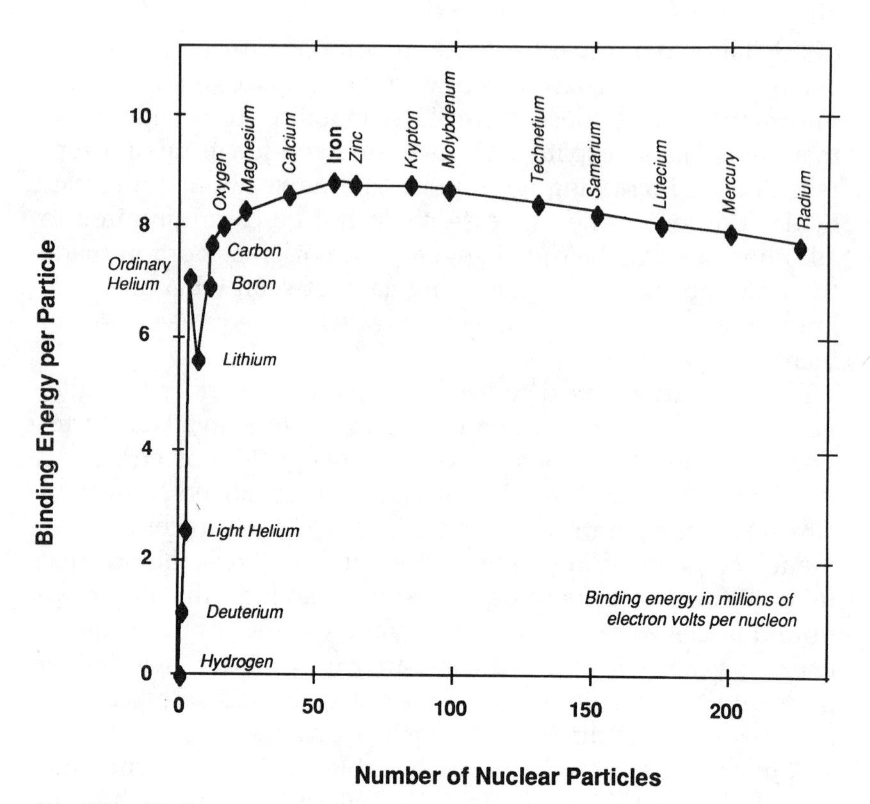

FIG. IV-5. Binding energy per nuclear particle as function of number of particles in nucleus. Represents surplus energy particles release when bound into nuclei by strong-force attractions. No additional energy can be extracted by fusing iron, so matter tends toward this least available state.

at the level of helium, leaving a long future in which energy was reserved in a potential mode. Expansion again permitted the entropy of the universe to increase, while at the same time the entropy content of nuclear matter was delayed from reaching its maximum value. Here and there in space, and now and again in time, that delay is interrupted when nuclear fusion reignites at the center of a star.

Once temperatures are measured in the ones, tens, or even

hundreds of thousands of degrees, the relevant physics shifts from nuclear to atomic. In atoms, the electrical attraction binding negative electrons to positive nuclei is so weak that collisions driven by temperatures of only some tens of thousands of Kelvin degrees can tear them apart. And several hundreds of thousands of years are needed for the universe to cool that much, so that electrically neutral atoms can persist. Once they can—an event that takes place when the universe's temperature is about 3,000° K—the material universe abruptly enters another age. The reason the new epoch is significantly different is that neutral atoms are inefficient scatterers of light. Your own observations of nature may already have told you so, for you can see a very long way—maybe as much as a hundred miles—on a clear day in a southwestern United States' desert, even though you are looking through an enormous number of *neutral*

FIG. IV-6. Monument Valley, Arizona. Light travels straight through the enormous number of neutral atoms in this desert's atmosphere.

Fig. IV-7 Spicules on solar surface. Light travels a tortuous path through the ionized plasma of the solar interior. Only that light scattered off the outermost layer of electrons is seen. (National Optical Astronomy Observatories.)

atoms residing in earth's atmosphere.[2] Your eyes also reveal the contrast with lines of sight through ionized plasma (the state of matter consisting of disassociated positive nuclei and negative electrons). The sun is a spherical collection of ionized plasma, and one does not see far into its interior. The light seen

[2] Every cubic centimeter of air contains about 3×10^{19} molecules. If you look through an area of 1 cm^2 for a distance of 100 miles, you are seeing right through 5×10^{26} molecules—five hundred trillion trillion of them!

is only that escaping at the surface. Inside the sun photons and unattached electrons ricochet off each other like billiard balls. A solar photon's path is consequently so zigged and zagged that it cannot be traced to its origin near the center, only to the site of its last scattering just beneath the surface. By the same token, when photons collide with particles of matter, they exert a pressure on them. It is this pressure that keeps the sun's outer layers from collapsing inward.

By analogy, early in the universe, when the matter was ionized (electrically charged), it was opaque to the passage of light. Photons and electrons bounced off each other, maintaining a close coupling between them. But when, after perhaps three-quarters of a million years, the matter switched to neutral (electrons combining with protons into atoms), matter and radiation were decoupled, each free to go its separate way. The universe switched from opaque to transparent. Matter, then, could begin to aggregate, since its gravity was no longer opposed by radiation pressure. Clusters, galaxies, stars, and other gravitational pockets collapsed into being.

Gravity can impose and maintain order (low entropy) in defiance of nature's tendency toward chaos (high entropy) only by compelling matter to fall toward some density concentration. In the process of falling the matter can do work, as it does when, for example, water falling from a dam spins a turbine. When the matter is uniformly spread over a sizeable volume, falling is long and slow; work can be extracted for a long time; accordingly, the initial configuration can be assigned low entropy. When, on the other hand, matter is tightly clumped, little infall remains possible, only a small amount of work can be extracted, and the system's initial entropy is high. Since even a few tight clumps can release useful energy by falling upon each other, the highest entropic state would have all matter tightly bunched in a single assemblage. How has this fate been avoided or delayed? The answer is that the initial explosion of space and time had just the proper vigor to fling matter so far apart that now the time required for it to collapse back to a point, its so-called free-fall time, exceeds the age of the universe. The big bang was such that matter avoided the dead-end state of maximum gravitational entropy if not forever, at least for eons and eons. Furthermore, as gravity began aggregating matter, thereby "cashing in" some of the available energy that expansion had placed in reserve, the matter's descent first had to overcome the delay imposed by thermonuclear fusion (as well as additional obstacles mentioned in the preceding chapter).

We have now seen examples of systems temporarily lowering their entropy by increasing by a more than compensatory amount the entropy in the larger whole to which they are openly connected. Living matter is no exception. Indeed, there are no free lunches for the hungry organisms on earth. The price for their existence is paid by the sun, whose nuclear entropy steadily increases via the conversion of hydrogen into helium. The order of the fuel at the solar center is dissipated into the disorder of heat and light, which seeks an equilibrium with the cold reservoir of interplanetary space. An energy gradient (or flow) results, and living systems can impose order on matter and energy here on earth by exploiting it. The spatial boundaries of the ecosystem integrating earth's biota therefore extend well beyond the planet itself. In addition, the ancestry of terrestrial life must be pushed back early enough in time to precede the sun's formation. Since this chapter reveals how conditions early in the universe's history influenced the possibility and then the rate of star formation, life traces its origins to the very beginning of the universe.

Life on earth is not a single type of organism or a fixed collection of types that remain the same throughout time. Instead, it is an interactive system with a continuity unbroken over its entire history. The continuity has progressed from simple to complex, from small to large, from monoculture to diversity, from instinctive to intelligent. Today's earth is a more intricate and complex ensemble than antiquity's simple world of almost exclusively blue-green bacteria. Does this progression not seem a violation of thermodynamics' second law? Not at all, when all the increments and decrements in entropy are assessed. Today's information-rich biocommunity built incrementally upon its predecessors. Of the innumerable mutations that preceded the emergence of humans, most—nearly all, in fact—failed the trial-and-error tests to which they were subjected. This is so because the living world, before longer-necked giraffes or longer-legged gazelles or whiter polar bears or more insect-attractive flowers, was already a finely tuned and coordinated enterprise capable of mastering its environmental challenges. Arbitrary changes were unlikely to be improvements. But at times of rapid environmental change or of major natural catastrophe, new opportunities were presented whose exploitation could perhaps be accomplished more fully by new variants of life. In this way, additional information (lower entropy) could accrue to the collectivity pulsing with the living process. But the far larger catalog of evolutionary failures represented wastage, lost opportunity, a decline in useful information, a degradation of the lifestream—in short, an increase in entropy. Life profited from failure. Exact duplication of already existing life forms would have been a prescription for disaster. In random chance lay opportunity.

Expansion of the universe is such a successful explanation for the emergence of order that one is tempted to invoke a (literally) prehistoric epoch of extremely rapid, even runaway, expansion as an aid to creating the type of universe we inhabit. Without such an epoch, we have no explanation for either the *flatness* or the *isotropy* of the universe.

Flatness refers to the fact that the mass density of the universe is close to the value separating continuously expanding (open) universes from eventually collapsing (closed) ones. In an open universe, space curves all the way to infinity without approaching any limit. An analogy in a two-dimensional space is the surface of a saddle. In a closed universe, space is so sharply curved that it curls back upon itself. The standard analogy in two dimensions is the surface of a balloon. Why, whatever the

FIG. IV-8. *The low entropy of the living opossum (left) has been more than paid for by the high entropy consumed in dinosaur extinction (right). (Dinosaur National Monument, Utah.)*

precise shape of the universe, should it lie so close to the flat space between these two opposite senses of curvature? The answer is that a universe that underwent *runaway inflation* would have overwhelmed most traces of curvature. When, for example, the balloon of our two-dimensional analogy is enormously distended, an observer anywhere on its surface has difficulty detecting curvature, but instead senses flatness.

The isotropy problem can also be explained by an inflationary phase in the universe's distant past. If every portion of the universe is to be the same and is to expand precisely such as to preserve that sameness, then all portions must once have been in such close contact that their differences could be erased. This could happen only if a small sample of space-time, a bubble in the ocean of totality, if you will, expanded ultrarapidly. Since the contents throughout all of the initially small sample would have been in almost instant communication with each other, there is no chance that different zones could expand without knowledge of how all others were behaving. In other words, anisotropies, or direction-dependent differences, could not accrue.

The notion that we occupy an enormously distended, but initially minute, bubble raises the question whether other bubbles percolated the froth of a primordial space-time. Perhaps there never has been a nothingness, but instead a continual belching of space-time domains of varying durations. Each domain, or bubble, could have prospered or expired at an arbitrary

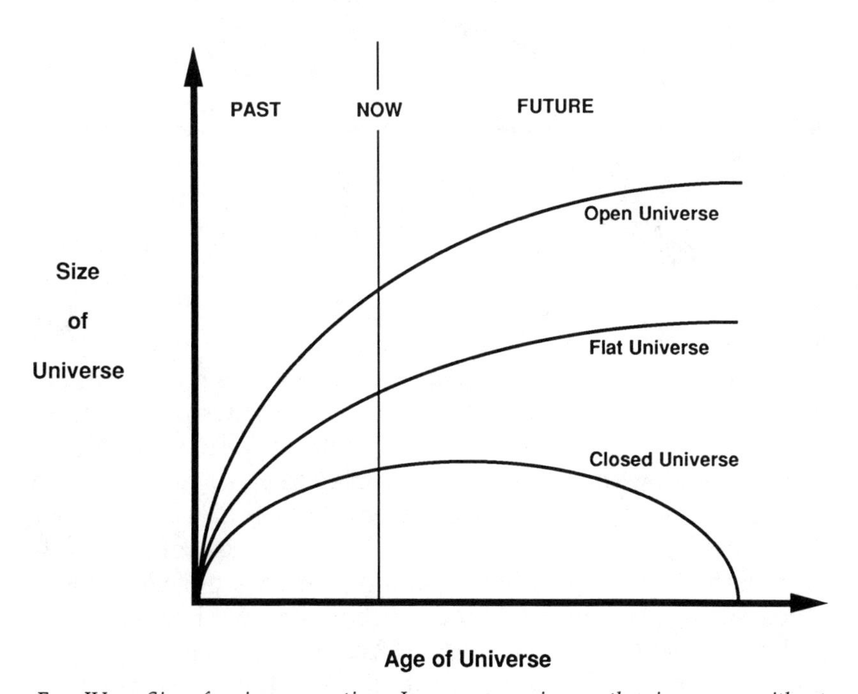

FIG. IV-9. Size of universe vs. time. In an open universe, the size grows without limit. In a closed universe, the size reaches a maximum, then collapses to zero. In a flat universe, infinite size is reached only after infinite time. Why should the universe be so close to this special case?

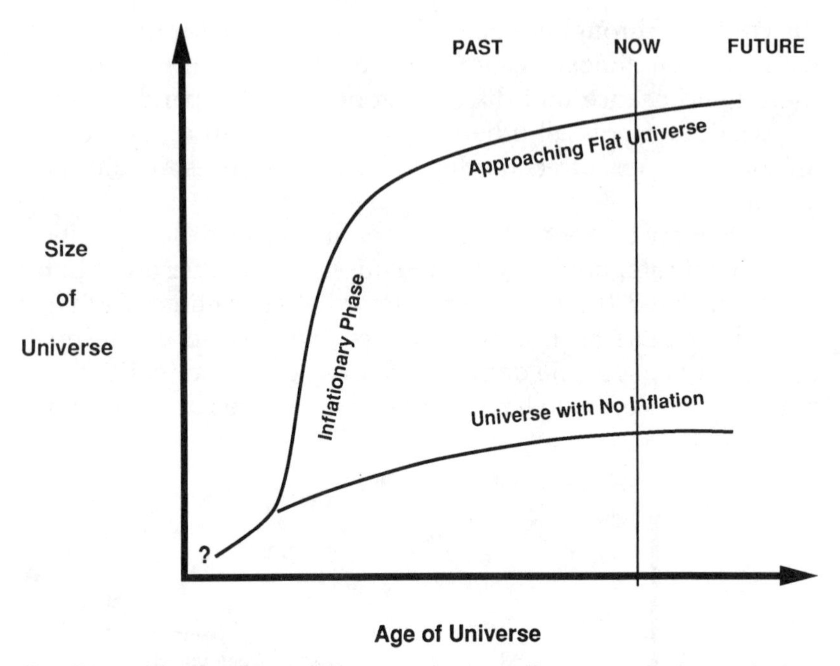

FIG. IV-10. The size of an inflationary universe swells enormously during the first fraction of a second after the big bang.

rate, but none could have contacted any other. If we regard each as a self-contained universe, then we achieve the penultimate mediocrity prescribed by the Copernican revolution. Whereas we once thought our planet unique and uniquely located, we now know of eight other planets in the solar system and suspect an enormous number elsewhere. Whereas our ancestors believed the sun unique, we now know of hundreds of billions (10^{11}) in our galaxy alone. Whereas our galaxy was once regarded as one of a kind, we now know it has hundreds of billions of likenesses elsewhere. What, then, is left for fragile human egos if our universe likewise is typical rather than special?

One answer might be humans' acute perspicacity. We can revel in the knowledge that we were able to discover our own mediocrity. That knowledge, or certainly the knowledge that we have been able to acquire it, should convey its own reward. The creative intelligence that was able to outline the foregoing scenario is itself a valuable property of the universe. We matter because we can know; but once we know, we learn that we hardly matter. We are creators, but also mere creatures. We can glimpse glory, perhaps even create it, but we are dwarfed by its magnificence. We are caught in the eternal paradox of being, on the one hand, minor participants in a universe that, on the other hand, without us might be nothing.

For Further Reading

Atkins, P. W. 1984. *The Second Law.* New York: Scientific American Books.

Campbell, Jeremy. 1982. *Grammatical Man: Information, Entropy, Language, and Life.* New York: Simon and Schuster.

Commoner, Barry. 1976. *The Poverty of Power,* pp. 6–29. New York: Bantam Books.

Crease, Robert P., and Mann, Charles C. 1984 August. How the universe works. *The Atlantic Monthly* 254(2): 66–93.

Davies, P. C. W. 1979. "Order and disorder in the universe." In *The Great Ideas Today.* 1979. ed. Mortimer J. Adler, pp. 44–55. Chicago: Encyclopedia Britannica, Inc.

Davies, P. C. W. 1978. *The Runaway Universe,* pp. 104–116. New York: Penguin Books.

Davies, P. C. W. 1986 September. The arrow of time. *Sky & Telescope* 72(3): 239–242.

Lake, George. 1984. Windows on a new cosmology. *Science* 224: 675–681.

Pagels, Heinz R. 1985. *Perfect Symmetry,* pp. 234–243; 331–352. New York: Bantam Books.

Schramm, David N. 1983 April. The early universe and high-energy physics. *Physics Today* 36(4): 27–33.

Smith, David H. 1983 March. The inflationary universe lives? *Sky & Telescope* 65(3): 207–210.

Trefil, James S. 1983 May. Closing in on creation. *Smithsonian* 14(2): 33–42.

Trefil, James S. 1983 May. Closing in on creation. *Smithsonian* 14(2): 33–42.

Trefil, James S. 1983. *The Moment of Creation*, pp. 158–170. New York: Macmillan Publishing Company.

Thomsen, Dietrick E. 1983 February 12. The new inflationary nothing universe. *Science News* 123(7): 108–109.

Waldrop, M. Mitchell. 1983. Inflation and the arrow of time. *Science* 219: 1416.

Waldrop, M. Mitchell. 1983. The new inflationary universe. *Science* 219: 375–377.

Waldrop, M. Mitchell. 1984. Before the beginning. *Science 84* 5(1): 44–51.

FIVE

Life, the Continuing Experiment

I am an experiment on the part of nature, a gamble within the unknown, perhaps for a new purpose, perhaps for nothing, and my only task is to allow this game on the part of the primeval depths to take its course, to feel its will within me and make it wholly mine.

Herman Hesse

THE EXISTENCE OF ORDER in a universe whose tendency is toward chaos, and in particular the maintenance of that order despite the passage of almost twenty billion years in which to dissipate it, is a remarkable legacy of the expansion born in the big bang origin of the universe. Nowhere that we know of with certainty (although it would be presumptuous, arrogant, and chauvinistic to exclude all other locales) is the superstructure of order as advanced as on the planet we inhabit. The order reveals itself, on a scale large enough to be noticed from a cosmic perspective, through certain of earth's properties, which are anomalous with respect to those of her nearest neighboring planets. Each anomaly represents a departure from equilibrium, and hence an improbable, nonrandom, low-entropy state. Especially provocative is the fact that the anomalous planet is also the one bearing life.

One telltale anomaly associated with the earth is its moderate temperature. Even more remarkable is that temperature's constancy over several eons. The theory of solar energy production and observations of sunlike stars at all stages of their development both indicate that the sun's luminosity has increased by a large factor—estimates range from thirty to one hundred percent—since it and the solar system came into existence. As a consequence, the earth has been exposed to a growing flux of radiation. Yet both the geologic and the biologic record (specifically the persistence of life on the planet) indicate that earth's mean surface temperature has remained close to the present 13°C for most of the planet's history. By analogy with the ability of humans to maintain a constant internal temperature of 37°C in an environment that fluctuates at least between 0° and 40°C, one is tempted to ascribe some of earth's thermal stability to the activity of the life that abounds there. It is certainly true that the chemical reactions that are the essence of the life process are nourished by the moderate thermal environment to which they are, and have been, exposed. The reactions are assisted particularly by the remarkable properties of water as a solvent, properties that exist only when water is in the liquid state. On Mars, whose mean surface temperature is close to −60°C, life would have been slow to evolve and, had it done so, would have had very slow metabolic rates. The surface of Venus, on the other hand, at 460°C is so hot that many molecules would have dissociated.

Another anomalous departure from equilibrium is exhibited by the chemical composition of earth's atmosphere. Neither the Venusian nor the Martian atmosphere contains appreciable molecular oxygen (O_2), whereas one-fifth of the molecules in Earth's atmosphere are O_2. In this instance we *know* that the source of the anomalous gas is life. Earth's initial atmosphere was virtually devoid of oxygen. But billions of years ago, distant ancestors of today's plants began using the energy of sunlight to manufacture carbohydrates from carbon dioxide (CO_2) and water (H_2O) in a process called photosynthesis. A byproduct of this process was oxygen, which was expired as a waste gas to the atmosphere. Its presence there, especially in association with other atmospheric constituents with which it could interact, is exactly the kind of disequilibrium, or local decrease of entropy, that labels earth an information-rich planet. Oxygen reacts so readily with other chemical elements that it is unusual to see such a high concentration of it as in Earth's atmosphere. If its concentration there increased only moderately to, say, about twenty-five percent of all atmospheric molecules, even trees in rain forests might *spontaneously* ignite. Something must account for this anomalous abundance of oxygen, and that something is life.

Chemical, physical, and biological disequilibria are not the only peculiarities remote observers of the earth might notice. They might also spot the existence of cycles, and cycles within

cycles within cycles, all interactive, circulating matter and energy on a variety of scales. In one such, the hydrologic cycle, water vapor in the atmosphere condenses into liquid droplets, which fall upon land so sculptured that they are funneled through lakes, rivers, and streams to the great ocean basins. From these the cycle is completed when solar energy evaporates water back into the atmosphere.

The patient observer would notice this hydrologic cycle intersecting a grand geologic one. In it the flowing water erodes sediments from surface rocks and carries them to burial grounds at the continents' edges. As layer after layer accumulate, their increasing weight metamorphoses the sediments into rocks or, at sufficient depths, even melts them into a magma whose extrusions become igneous rocks. Deformation and uplifting return sedimentary, metamorphic, and igneous rocks to the surface, where they again lie exposed to weathering and erosion.

How, one might ask, are the deformations and uplifts accomplished? The answer invokes another cycle, this one within earth's molten interior. There, convection currents drive the material upon which the continents ride. When continents collide, or when the sea floor slips under them, land masses are thrust upward, producing changes necessary for the geologic cycle. Other forces of change include volcanism, which transports material from earth's interior to its surface and its atmosphere (including water vapor, which enters the hydrologic cycle), and its inverse, subduction, by which solid material is returned to the interior at places where one land mass is driven beneath another.

All such cycles signify dynamism. Their interconnectedness

FIG. V-1. (Top) Earth supports a rich and varied biota, because the temperature at its surface is conducive to an active chemistry. (Middle) Martian life is nonexistent or exceedingly sparse, a consequence of the slow rate of chemical reactions in the planet's frigid environment (Photo NASA/JPL). (Bottom pair) Venus is devoid of life because its climate is so hot no molecules can be preserved (Soviet Venera 13 and 14).

FIG. V-2. The Grand Canyon of Yellowstone National Park, Wyoming, USA. The Yellowstone River is a small part of earth's hydrologic cycle, while the canyon it has carved is a product of a geologic cycle.

hints at a unity of function, an intermeshing of processes whose synthesis is no longer a straightforward sum of individual occurrences. Nowhere are these unifying and integrating syntheses more tightly woven than in the cycles looping through living systems. The fact that mammalian blood, or cytoplasm in general, bears a close chemical resemblance to sea water testifies to the importance of the hydrologic cycle for living organisms. Water provided a medium within which, early in earth's history, the ingredients of life-sustaining molecules could be suspended, transported, and aggregated into concentrations in which linkages among them could occur.

Another cycle can be seen if we follow a single carbon atom over a considerable period of time. It might start as a link in a sugar molecule resident in a cell of a mammal. When, after some decades, the mammal dies, bacteria begin decomposing its flesh. A particular bacterium ingests the sugar molecule we have identified, and later uses the energy stored within it to power its own activities. To utilize the energy, it burns the sugar molecule, expiring the carbon atom of interest to the atmosphere as part of a molecule of carbon dioxide. There it may remain for centuries until, perhaps far from its release point, a plant absorbs the CO_2 molecule and fixes the carbon atom into a leaf. The leaf may some day be eaten by an insect, the insect by a bird, and the bird by a mammal, returning the carbon atom to the setting in which it started. Had it followed an alternative route, the flagged CO_2 molecule might have been absorbed into the ocean, become incorporated into a carbonate, and later deposited out as a limestone. Perhaps millions of years would have transpired before erosion returned the original carbon atom to the life cycle. Either way, the cycles illustrate that *the separation of matter into animate and inanimate components is temporary*. Life and non-life interpenetrate. Life and the land, sea, and air in which it exists are inseparable. A quantitative estimate suggests that a third of all chemical elements are at some time recycled biologically. An oxygen molecule in the atmosphere is, on the average, part of a living system for about one day of every decade. A hydrogen atom in ocean water spends about one minute of every year in a living organism.

Life, then, is an integral part of the planet; the physiography of the earth is incomplete unless life is included as part of it. The living part of the planet is not ordinary, though: it contains the most highly ordered structures among the many that exist. Because of this high degree of ordering, life's various manifestations are correspondingly improbable. How improbable? Consider the case of a single human.[1] The genes humans inherit from their parents determine their physiological characteristics—sex, height, weight, color of skin, hair, eyes, and so forth. Each human gene consists of some four billion (4×10^9) units

[1] The probability calculation is from I. S. Shklovskii's and Carl Sagan's *Intelligent Life in the Universe* (New York: Dell Publishing, 1966), p. 196.

FIG. V-3. A strawberry coral, a highly ordered configuration of chemical elements that is extremely improbable because it departs so enormously from a random configuration.

assembled in a prescribed sequence. Each unit can be one of only four specific molecules. It follows that one of four possibilities can occupy the first position in the genetic chain. Adding the second link offers $4 \times 4 = 4^2 = 16$ total possible configurations; the third link presents $4 \times 4 \times 4 = 4^3 = 64$; and so on, until completing the chain presents the staggering total of $4^{4 \times 10^9} = 10^{2,400,000,000}$ possibilities, among which each of us is but one. The number of possible configurations is far greater than the number of particles in the universe!

If one had thrown four billion four-sided, pyramid-shaped dice every second since the universe began (for a total of less than 10^{18} throws), the chance that he would have happened upon the sequence of his own genetic composition is absolutely infinitesimal. Then how can we possibly be? Obviously we cannot have sprung full-blown from a pre-existing state of chaos in which all the atoms in our body were scattered hither and yon. We can only have arisen slowly, piece by piece. We must have been preceded by several stages of aggregation of matter into, first, simple chemical units, then into more complicated combinations of these simple units, followed by combinations of these combinations, and so on, until the full glorious complexity of a human gene was assembled. Do we share a common ancestry with the rest of earth's life, and hence represent the patient, steady accumulation of the information that describes us? The answer is overwhelmingly affirmative.

At this point, we can begin to appreciate the web in which we humans are enmeshed. It extends in two orthogonal directions: (1) "vertically," so to speak, through time—in the way, just mentioned, in which small bits of information have been integrated into successively larger ones; and (2) "horizontally," in each epoch, through all the biota in existence then. The evidence that all earth's biota are woven into a common web manifests itself in the way genetic information is stored, a point mentioned as early as the first chapter, but one that warrants additional consideration. The storage of genetic information is the same whether the organism is single- or multi-celled, whether it is plant or animal, ancient or modern, whether it walks and talks or is stationary and silent. The genes in all these cases are immensely long molecules called nucleic acids. DNA, for *d*eoxyribo*n*ucleic *a*cid, is the shorthand representation of the substance that determines the particular form in which a living organism appears. Every cell of that organism contains closely similar DNA molecules. Those molecules instruct the organism how and when to make more of the ones it needs to stay alive and to reproduce. DNA itself is constructed from just four identifiable molecular units called nucleotides and labeled, again in shorthand, A, C, T, and G. The order in which these four nu-

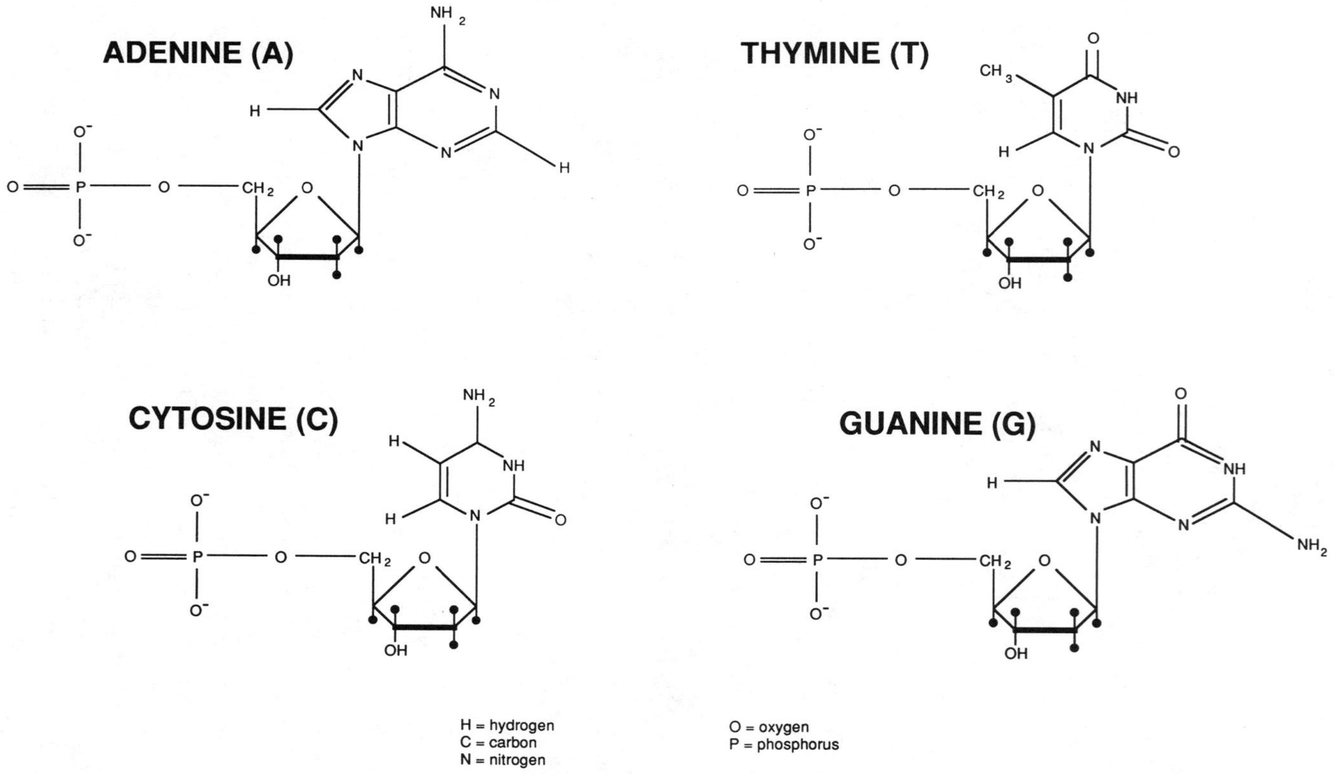

ADENINE (A)

THYMINE (T)

CYTOSINE (C)

GUANINE (G)

H = hydrogen
C = carbon
N = nitrogen

O = oxygen
P = phosphorus

FIG. V-4. *The four nucleotides, which, in various sequences, specify the DNA of every living thing.*

cleotides are arrayed and the length of the full sequence together determine the nature of the organism. For a bacterium, about two million nucleotides must be assembled in a specified order, while a human requires some four billion. The former is approximately equivalent to the information content of twenty average books, and the latter to some tens of thousands of books.[2] That each "book" (or organism) is "written" in the same language (of DNA) based upon the same alphabet (the four nucleotides) is decisive proof linking bacteria, humans, and the full spectrum of life between.

It is not just the genetic material we and rhododendrons, for instance, pass to our descendants that is similarly constructed, but also the substances of which we and they consist. The critical "stuff of life," regardless of which organism we consider, is protein. Every cell of every organism contains thousands of different proteins. These govern the rate of the chemical reactions that determine an organism's metabolism. If proteins were not present to speed certain reactions, biology would slow to the pace of ordinary chemistry; chaos would conquer order; life would cease; and the planet Earth would be about as exciting as its neighbor Mars.

Proteins also are extremely long molecules, albeit not nearly as long as DNA. They, too, are composed of subunits whose sequence determines which particular protein is constructed, and again the inventory of subunits is limited, this time to twenty varieties, each called an amino acid. Instructions encoded in the DNA tell how many, and in what order, amino acids are to be used. These instructions specify the protein. The method of protein construction is the same for azaleas as it is for zebras. Evidently creating life means putting links on a chain in a precisely predefined order. Furthermore, there is something special about the amino acids used by life that is the same

[2] Strictly speaking, this comparison is an exaggeration. Most (ninety percent?) of the human genome does not code for anything. Thus most of the human "book" is nonsense.

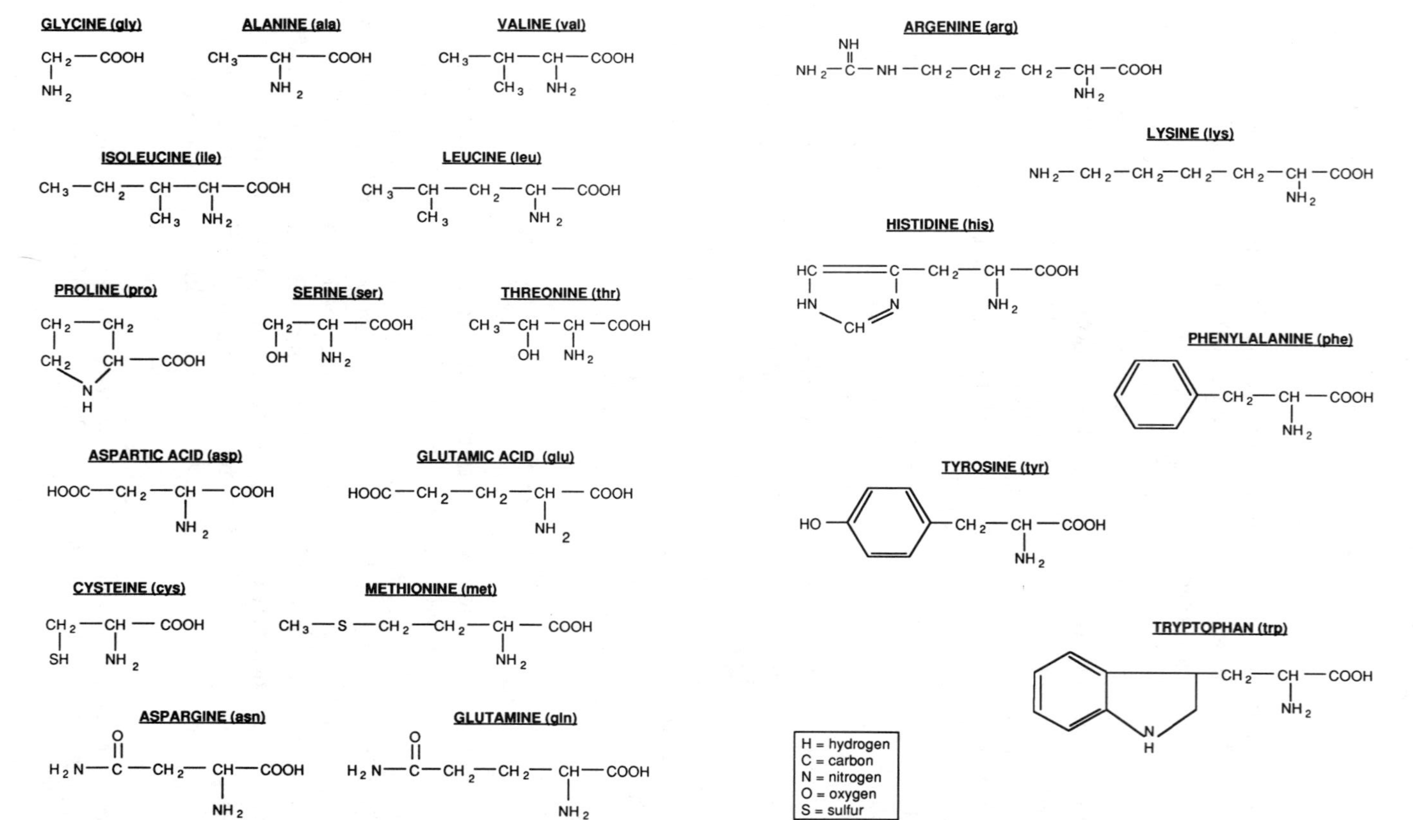

FIG. V-5. *These twenty amino acids suffice to make every protein found in all life forms on earth.*

for all specimens: the acid molecules essential to it are asymmetric. They could in principle appear in either of two forms, which are mirror images of each other (cf. Figure I–8). In fact, when proteins are synthesized in laboratories, equal numbers of the two types are produced. Even in meteorites, molecules with both "handednesses" appear. But among all the living organisms on earth only the left-handed variety is utilized, additional testimony to their biochemical relatedness.

Another feature common to all earth's organisms is the modular method of construction each uses. The basic module is the cell. Nothing lives that is simpler than a cell, and nothing more complex lives that was not first a single cell.[3] All organisms therefore control the entry and exit of materials through cell boundaries. Within cells, all use identical molecules to serve as the carriers and storers of the energy needed to maintain the cell's internal order.[4]

All of the foregoing similarities and identities reflect the enormous degree of ordering that distinguishes our planet from its neighbors. In this context we mean by order that which is distinguished from randomness or chaos. A commonly used example in literature is the output of monkeys at typewriters. From the total population of random strings of letters, spaces, and punctuation, only a minute fraction, the ordered, has any meaning or useful information content. Life, too, represents a selective narrowing of options. For example, given the hundred or so atomic elements that exist here on earth, an enormous variety of simple organic molecules is conceivable. But not all are represented in earth's biota. Instead of a few examples each of a huge number of molecules, there has been a natural selection of very large numbers of just a few kinds of molecules.

Order is also evident in the way DNA specifies "only" some

[3] Viruses are at the boundary between living and non-living. They can only function and reproduce (i.e., act alive) when inside a host cell. In isolation, therefore, they are dead.

[4] Adenosine triphosphate (ATP) is the name of the molecule providing the major source of chemical energy used in the metabolism of all cells of all organisms.

one and a half million types of animals and a half million types of plants. While at first glance these numbers seem large, consider that DNA is constructed from millions and billions of subunits, the various possible combinations and permutations of which swamp the insignificant fraction that is represented among the living. (Earlier in this chapter we calculated the possible varieties of human being to be $10^{2,400,000,000}$. Against this number, 10^6 or 10^7 is close to nothing.)

All of this, indeed the entire history of life on Earth, suggests that order begets more order. That is to say, three defining characteristics of living systems are the abilities (1) to create more living systems from non-living components (reproduction), (2) to mutate freely and then to reproduce these mutations (imperfectibility), and (3) to incorporate organic components for fuel and for repair of broken tissue (metabolism). The first two processes assure survival over the long run, and the third over the short. All assimilate high-entropic components into low-entropic systems. To do so requires a steady supply of energy of sufficient quality (i.e., with low enough entropy) that life can use it to perform useful work before discarding a less useful (higher entropic) form of energy to the environment. Because of this requirement, living systems have become particularly adept at capturing, storing, and transmitting energy. The steady flow of energy they tap into arrives from the sun.

The sun drives the process of *photosynthesis*, which constitutes the foundation of life's energy pyramid. Using photosynthesis, plants (and some microbes) are able to capture the radiant energy of sunlight and to store it as chemical energy in carbohydrates—molecules made from atoms of carbon, hydrogen, and oxygen. The sources of these atoms are carbon dioxide (CO_2), abundant in the atmosphere as it has been since earth's formative stages, and water (H_2O), likewise a major perennial constituent of earth's environment. When the carbon dioxide and water are synthesized into a carbohydrate, a molecule of oxygen (O_2) is left over. If and when the chemical bonds holding the carbohydrate molecule together are broken, energy is

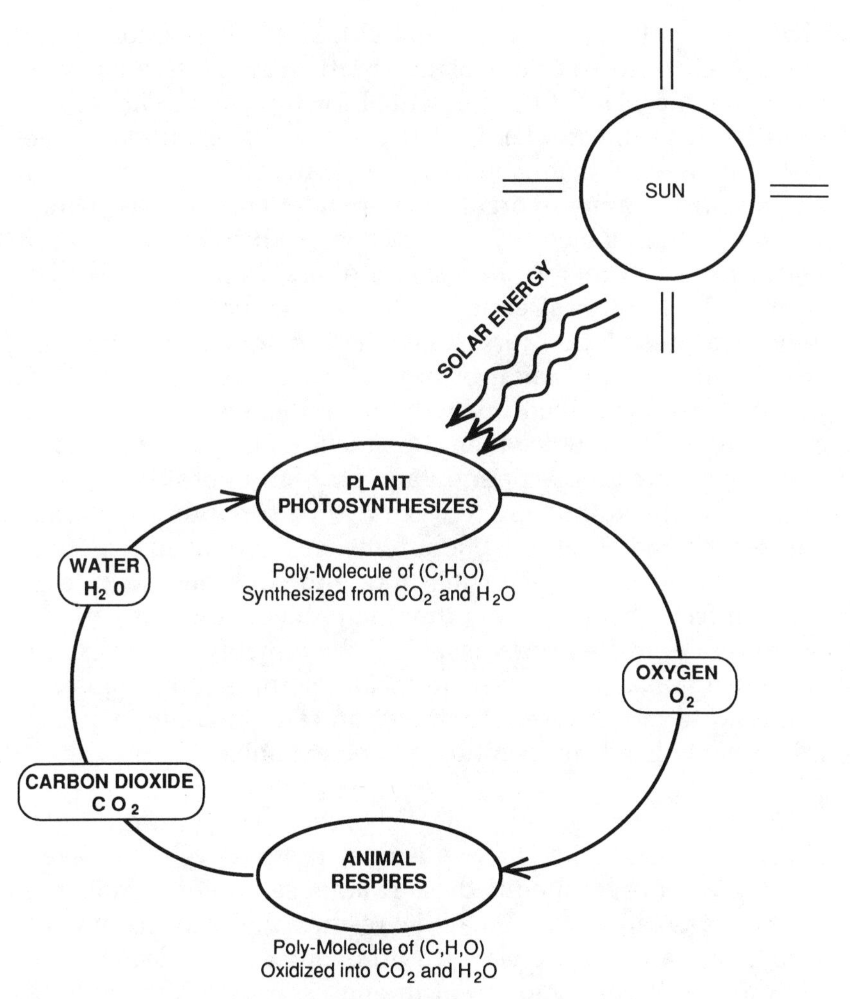

Fig. V-6. Photosynthesis relies on respiration's by-products. Respiration relies on photosynthesis's by-products. Solar energy drives the cycle.

released. The main (although not the only) method for breaking the bonds is *respiration,* a near inverse to photosynthesis. The plant itself, or some animal that has eaten it precisely for the purpose of obtaining the energy stored therein, respires by oxidizing (crudely, burning) the molecular fuel. In addition to the energy released, carbon dioxide is left over. Thus what was waste for the plant, O_2, is essential for the animal; its waste, CO_2, is, in turn, essential for the plant. Another strand in the web linking earth's inhabitants is woven. Since exhaustion of any resource dooms an organism dependent upon it, recycling, whereby "complementary" organisms produce resources necessary for each other, is a preserver of life. In other words, on earth and wherever else life exists, a single breed of organism may be a prescription for extinction. Continuity demands diversification for provision of mutual interdependencies. And the cooperation symbolized by these interdependencies is more crucial to life's existence than the much-heralded competition.

We have stressed a common ancestry, an unbroken thread, uniting all life with its predecessors. Such continuity implies a faithful scheme of reproduction. Now we have called attention to the advantages of diversification, and consequently a less-than-perfect scheme of replication, in reducing the competition for resources made nonrenewable if the only organisms extant are consumers of it. Because mutability permits continuous evolution into newer forms, competition can gradually be made less head-to-head; cooperation can be enhanced.

Perhaps a summary and some elaboration are in order. Earth hosts a presence called life, which has progressively increased the degree of order it embodies. To impose order life must exploit external energy and material resources, the availability of which vary around the planet. Within any one particular zone of availability, consumption of the necessarily finite supply of resources by only a single kind of organism would limit the number of those organisms that could exist. Furthermore, it would restrict the efficiency with which resources were utilized, since no single converter of environmental disorder into metabolizing and reproducing order can, in just one step, bring about the maximum lessening of entropy. Life has evaded these limitations and restrictions by evolving into a variety of forms,

TABLE V-1. *Size Distribution of Land Animal Species (From Harold Morowitz, 1982)*

Length of Animal	Number of Species
Less than 0.01 inches	20,000
0.01 to 0.1 inches	220,000
0.1 to 1.0 inches	600,000
1.0 to 10.0 inches	20,000
10.0 to 100.0 inches	1,500
Greater than 100 inches	10

each capable of utilizing different zones of availability, or of consuming from the same zone of availability by using different, noncompetitive techniques.

We see these principles in many manifestations, of which we shall mention only four. The first is the *range of sizes* over which earth's organisms are arrayed. In the second chapter, this was given as seven orders of magnitude, or factors of ten, using a fly as the midrange reference. This choice of reference was appropriate, since there are more fly-size animals than animals of any other size. The reason lies in the overwhelming numerical superiority of energy sources (food) that are about the same size as these animals, some few centimeters or so. Every animal needs to consume enough food to mend its body tissue. If an animal is small, it has little tissue, so its food consists of small items. These abound: every plant, for example, offering a multiplicity of leaves, branches, and roots. Large animals, on the other hand, have so much tissue they need a great deal of food; even though that food can be small items, there must be many of them. To find the necessary quantity a huge niche-space must be searched, one that cannot support an enormous number of these voracious animals. No wonder large species are rare, and intermediate-sized ones common.

But why no infinitesimally small ones? Here the answer lies

Fig. V-7. Whale migrating from Baja California to the Bering Sea. To feed its immense bulk, it must traverse enormous oceanic volumes.

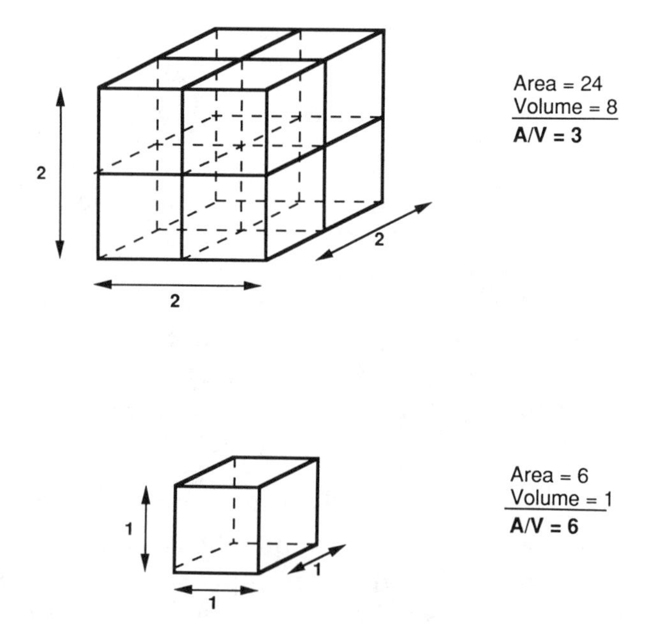

Fig. V-8. As the dimensions of any shape shrink, the ratio of surface area to volume grows at the same rate.

in the volume-to-surface inequality. If the dimensions of an organism shrink, its volume decreases faster than its surface area. If the height, width, and length of a specimen are cut in half, its surface area falls to one-fourth its original size, but its volume falls to one-eighth. For the very smallest organisms, the surface per unit of volume reaches its extreme. Consequently, the chemical substances inside these organisms have maximum exposure to the external environment. If this environment is air, dehydration is a constant threat because of the relatively large area through which the small quantity of water enclosed by the organism can evaporate. Most of the tiniest organisms must therefore occupy aquatic environments; some, of course, can survive in the soil or inside other organisms, but usually only where the surroundings are moist. Large bodies of water, though, are rather uniform niches. Their acidity, salinity, and other chemical features are fairly constant over immense regions. One therefore expects, and indeed finds, that they host few varieties of the smallest organisms, but many individuals of each variety.

The second obvious manifestation of life's effort to maximize energy utilization in its quest for higher order is its *spatial heterogeneity*. Life exists everywhere on the planet—ocean bottoms, mountain tops, arid deserts, polar ice caps, acidic hot springs. Indeed, the outer limits defining the boundaries of "livability" have probably not yet been determined. Life is not equally abundant everywhere, however. It is strongly concentrated at the interfaces where land, air, and water meet. Even a mere few meters from these interfaces—up into the atmosphere, or down into the soil or the sea—produce a sharp drop in the numbers and varieties of organisms present. The limiting factor in the downward direction is the extinction of energy sources. Sunlight, the prime source, can scarcely penetrate soil at all and water only somewhat better. Some organisms can exist beyond its direct reach by scavenging "tidbits" dropped by surface organisms or by synthesizing energy-rich structures from chemicals emanating from the ocean's floor—but only a relative few. In the atmosphere, the limit is set by gravity. A body of all but the most negligible size must expend energy continuously to stay aloft. The higher in the atmosphere it resides, the less air exists to buoy it up, so the more energy it must expend. The diminished supply of carbon dioxide and oxygen at high elevations makes energy production more difficult there, precisely where it is most necessary.

The abundance of living matter varies horizontally around the earth's surface as well as perpendicularly to it. Polar regions, relatively deprived of direct sunlight, are more sparsely populated than the energy-rich midlatitude and equatorial regions. The latter likewise offer a diversity of environments that can support varying quantities of biomass. Despite geographical fluctuations in abundance, life's most remarkable characteristic remains the diversity of structures into which it has branched so that it can utilize whatever quantity, wherever located, of energy and material available. Little potential niche space lies unexploited.

In the same way that spatial heterogeneity increases the range of occupancy of an environment, *temporal heterogeneity* increases its long-term stability. By this we mean the shifting composition of a biotic community as its physical environment undergoes changes. The most striking examples occur after natural catastrophes—volcanic eruptions, for example (see Figure I–1). First, organisms (bacteria, protists, and plants) capable of synthesizing their own food sources (called autotrophs) enter the environment rendered sterile by such catastrophes. These can be followed by primary heterotrophs (e.g., herbivorous animals) capable of deriving their energy by eating autotrophs. Finally, conditions become suitable for secondary heterotrophs, carnivorous organisms that can eat the primary heterotrophs. At different times in this sequence, different organisms dom-

→

FIG. V-9. Earth's environmental offerings are heterogeneous. The (top) arctic of Greenland supports far less life than the (bottom) Hoh Rain Forest in Olympic National Park, Washington.

inate. Eventually, a steady state is reached in which the organisms achieve the balance proper to sustain them all—until the next major perturbation occurs.

The existence of interacting *trophic levels* demonstrates how efficiently life utilizes solar energy to increase its internal order. There is an analogy between the way multicellularity, the cooperative and coordinated functioning of an assemblage of cells, enhances this efficiency and the way whole communities of individual organisms coordinate their actions toward the same end. That is not to say that a member of a community, or ecosystem, is aware it is contributing to the operation of the whole, any more than a single cell knows it participates in a larger unity. Rather, each structure—cell or organism—is part of the environment of every other structure in its unit; and each unit—organism or ecosystem—exists only by virtue of the entropy exchange it negotiates with the environment of which it is a part. Every living thing contains interacting units and is itself a unit in an interacting system.

What, then, is the proper unit for specifying life? Ecosystems are not comprehensive enough, for the operation of each ecosystem is dependent upon every other, all being linked through the great common reservoirs of Earth's atmosphere, hydrosphere, and lithosphere. Only the thin film girdling the entire Earth—the entity called the biosphere—is comprehensive enough to include all of the living process. The mass of this biosphere is approximately a billion billion (10^{18}) tons, most of which is ocean water. At any moment, the biomass, the living component, is but a tiny fraction of the whole. But the parts that constitute the instantaneous biomass are continuously recycled through the much more massive pool of nonliving material. The biosphere operates as a single patterned fabric; an intricately interwoven system of complex subsystems; a huge metabolic device for the capture, storage, and transfer of energy; a superorganism whose internal order is not only maintained, but gradually increased, by the energy that winds tortuously through it. *Life can be defined as the total activity of this superorganism.* Even the boundaries of life are flexible, since human representatives of the biosphere are contemplating spreading elsewhere the order that exists within it, and have already reconnoitered beyond earth's confines in an initial step toward that objective.

Clearly, only a cosmic perspective permits one to consider earth a single living engine. Only a retreat into space gave us a wide enough view of the web in which we were trapped to see that it bound all life together. Likewise, only a cosmic calendar equipped us to comprehend metabolic rates measured in millions and billions of years, the temporal yardsticks of the biosphere. Analogies may prove helpful. Organs within organisms are analogous to ecosystems within the biosphere. The cells composing organs resemble organisms constituting an ecosystem. The single cell from which an individual originates is matched by the single cell that was the original progenitor of all earth's life. The diversification of life from bacteria into fungi, animals, and plants illustrates the same strategy of maximizing energy utilization as the diversification in the biosphere from uniform ecosystems into varied ones (as from ocean to land, for example). The birth and death of individual organism is analogous to the origin and extinction of entire species in a superorganism.

One epochal development—a most recent one at that—has no analogy: the introduction of a community of organisms equipped to recognize their participation in a system bigger than they. Prior to human occupation, which means for nearly all of the universe's history, units—say cells in organisms, or individuals in animal societies—had no awareness of their contribution to a greater whole, even though their activities were constrained by that fact. Organisms in particular did not evolve, adapt, and diversify because they saw greater opportunity. Instead, as they reproduced, chance variations arose among them—inevitable consequences of nature's imperfectibility ex-

Fig. V-10. From a cosmic perspective, earth's biosphere reveals itself as a single superorganism. (NASA.)

pressed in the second law of thermodynamics. Enough variants were created so that some were suited for whatever new circumstances a changing world introduced. This was the sense in which life could be considered an experiment—but not by the organisms themselves. They just existed, and whatever happened, happened, and quite naturally, if the happening involved change, those who were most fit for the new realities survived it in greatest numbers. Until man—clever man. He deduced that, in fact, life had multiplied and grown willy-nilly; that the organisms that had paved the way for his existence had behaved like the chemicals in a chemist's experiment, passive ingredients in a process not of their choosing. Having once learned the rules of the game, humans no longer needed to be mere pawns in it. Their panevolutionary vision offered them the chance to manipulate the game to their own advantage.

Humans' newfound abilities were early applied to keep their weakest, sickest members alive longer than unmanipulated nature would have permitted under its strict dictum of "survival of the fittest." This immensely humane act was not balanced, however, by a reduction in birth rate equal to that in the death rate. As a result, human numbers burgeoned. Fortunately, this departure from the steady state can be corrected. Indeed, there are signs that already the birth rate is being decreased to match the death rate more closely.

Other actions of humans exhibit an ability—good or bad—to be active participants, not spectators, in the experiment of life. No other organism has become as powerful at manipulating the entropy interchanges separating order from disorder. At the input end, where energy is imbibed with a potential for

FIG. V-11. People in a dense gathering. Their numbers shot up rapidly when medical advances reduced the death rate without changing the birth rate.

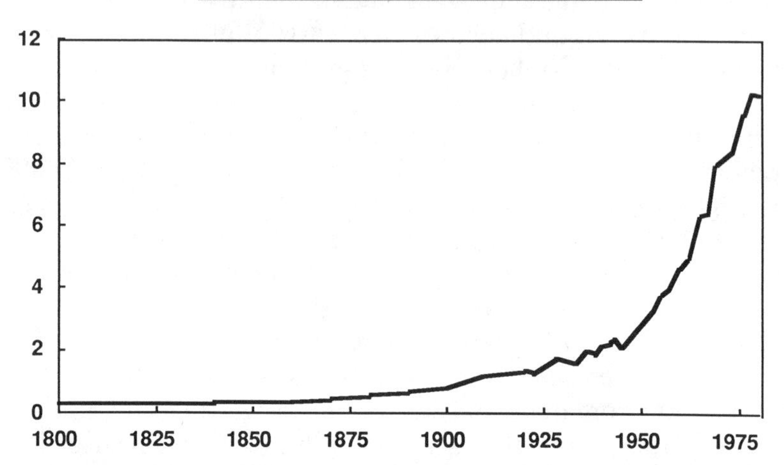

Fig. V-12. The history of humans' use of energy. (Data from Council on Environmental Quality 1981.)

increasing order, no creature has approached humans' ability to capture energy sources. Human history of the last two centuries effectively details the progressive utilization of ever more energy to fuel processes of his own design. Much of the order that this voracious energy consumption fueled is in the form of consumer items—vehicles, kitchen gadgets, bric-a-brac—and weapons, both of whose value has extremely limited duration. Also, the entropy increase delivered to the environment at the output end of every process necessarily sows greater disorder than the amount by which the process decreased it. These discards are now having macroscopic effects, effects noticed on a biospheric scale. Atmospheric testing of nuclear weapons during the 1950s spread radioactive waste around the planet. Exponentially growing consumption of fossil fuels increases the carbon dioxide content of the planetwide atmosphere, contributing to a global warming with incompletely understood consequences. Deforestation in the tropics likewise adds carbon dioxide to the biosphere by eliminating sources

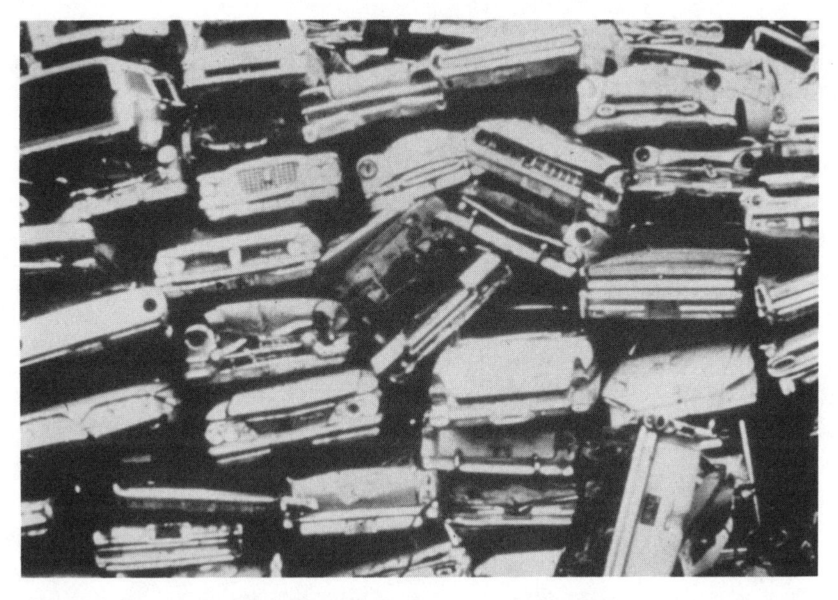

FIG. V-13. Order was imposed via an enormous utilization of energy, but only temporarily. Disorder, is now being returned to the environment in an amount greater than the manufacture of these automobiles had increased order.

that consume it. Chemical byproducts of industrial processes introduce acids to the atmosphere and hydrosphere—not just locally at the source, but over major portions of the globe.

None of these human-directed activities, nor the many others we could have mentioned, need be carried on exactly as they are today. Most are examples of an ability that no other creatures on this planet possess and that humans themselves acquired only a blink of an eye ago in earth's history. One should not conclude that the ability of humans to direct the experiment of using external energy to create internal order, while simultaneously participating in that experiment, is necessarily bad. Its potential for doing good is as great as that for destruction. In fact, directed manipulation of life's order-giving ability will in the future be absolutely essential to its continuity (see Chapter IX). But there is no escaping the fact that the price we pay for the power we have gained is a great responsibility. We alone can decide whether the experiment that created us naturally is worth continuing consciously. Life here will never again enjoy the luxury of unhindered development. Henceforth it will bear the stamp of human intervention—for better or for worse.

For Further Reading

Colinvaux, Paul. 1976. *Why Big Fierce Animals are Rare*. Princeton, NJ: Princeton University Press.

del Morel, Roger. 1981 May. Life returns to Mount St. Helens. *Natural History* 90(5): 36–47.

Feinberg, Gerald, and Shapiro, Robert. 1980. *Life Beyond Earth*. New York: William Morrow and Company.

Grobstein, Clifford. 1974. *The Strategy of Life*. San Francisco: W. H. Freeman and Company.

Hoagland, Mahlon B. 1978. *The Roots of Life*. Boston: Houghton Mifflin Company.

Levi, Primo. 1984 September-October. Travels with C. *The Sciences* 24 (5): 16–21.

Lovelock, James, E., and Margulis, Lynn. 1974. Atmospheric homeostasis by and for the biosphere: The Gaia hypothesis. *Tellus* 26: 2–9.

MacMahon, James A. 1982 May. Mount St. Helens revisited. *Natural History* 91(5): 14–24.

Morowitz, Harold. 1982 November. Beetles, ecologists, and flies. *Science 82* 3(9): 28–30.

Seiver, Raymond; Jeanloz, Raymond; McKenzie, D. P.; Francheteau, Jean; Burchfiel, B. Clark; Brocker, Wallace S.; Ingersoll, Andrew P.; and Cloud, Preston. 1983 September. The dynamic earth. *Scientific American* 249(3): 46–189.

Shklovskii, I. S., and Sagan, Carl. 1966. *Intelligent Life in the Universe*. Translated by Paula Fern. pp. 182–200. New York: Dell Publishing Company.

Toon, Owen B., and Olson, Steve. 1985 October. The warm earth. *Science 85* 6(8): 50–57.

Weimer, Jonathan. 1986 July/August. In Gaia's garden. *The Sciences* 26 (4): 2–5.

SIX

Earth as a Superorganism

There is nothing inorganic. . . . The earth is not a mere fragment of dead history, stratum upon stratum, like the leaves of a book, to be studied by geologists and antiquaries chiefly, but living poetry like the leaves of a tree, which precede flowers and fruit; not a fossil earth but a living earth. . . .

Henry David Thoreau

PHYSICAL STRUCTURES IN THE UNIVERSE are organized into a hierarchy of order. Life, too, is characterized by such a hierarchy, webs within webs within webs of structure and functional control. The cosmic perspective this book is developing permits us to see the entire biosphere as a single living unit, a superorganism. It is true that parts of it can be distinguished as different, although not independent, ecosystems: each ecosystem can be subdivided according to mode of energy consumption, either autotrophic or heterotrophic, and further, both of these modes of energy consumption take place via whole communities of living organisms, some (e.g., insect societies) so integrated that they function as single units. We also know that each such community is populated by individual organisms, each divisible into organs, themselves divisible into cells, within which are organelles encasing macromolecules synthesized from simple molecules. These molecules, in turn, are strings of atoms; and on and on into the subatomic and subnuclear realms. The whole decomposition of life into ever smaller components resembles a Russian doll, within which is another, surrounding yet another, and so on to some final miniature figure. But the resemblance is only superficial; there is a fundamental difference. The levels of the hierarchy of life are not distinct from each other; instead, each is made of units from the level below and helps constitute a part of the level above. Nor is any unit of a certain level independent from the other units on that level. Biological, chemical, physical, and geological processes join all parts of the hierarchy to all others. In other words, the hierarchy is a convenient way for people to characterize life, to give them a framework for describing it. But this convenience for humans does not mean nature adheres

Fig. VI-1. Nesting Russian dolls.

to the same simplifying strategy. The hierarchy of life collapses if any one level is removed from it.

The interrelatedness of all life is strikingly illustrated by some exceptionally strong interdependencies called symbioses. Insects and flowers have coevolved some of the most intricate of these reciprocally beneficial relationships. In one well-studied example, bumblebees seek to extract the maximum amount of energy stored in the nectar of delphiniums. To do so, the bees always start near the base of an inflorescence and work vertically upward. The flowers they encounter at the bottom are older, bigger, and contain more nectar, so a bee is "mining" the most rewarding energy source first. By traveling straight up, it misses some flowers, since they are arranged in a spiral pattern, but energetically its path makes sense. Had the bee instead chosen to visit every flower on the stem, it would have expended considerable energy in sideways motion, more, in fact, than it would have gained back as it rose to the flowers

Fig. VI-2. Praying mantis on rose. Insects and flowering plants have co-evolved some intricate symbioses. (Photo: Carl A. Seielstad.)

with less nectar. At some point, more nectar per second can be acquired by leaving the top of one plant for the bottom of the next. Advantages accrue not just to the bee, but also to the plant. For the delphinium's flowers are segregated sexually: the larger, older, nectar-rich flowers at the bottom of an inflorescence are female, while the younger ones at the top are male. Consequently a visiting bee arrives at a female flower immediately after having visited a male of a different plant. Efficient fertilization is assured. The plant's spiral arrangement of flowers itself is the result of intimate coevolution with bumblebees. The geometry discourages a bee from satiating itself at a single plant. If it did so, it would fail to deliver pollen to a neighbor.

Both delphiniums and bumblebees have options besides reliance upon each other. Bumblebees can acquire nectar from other kinds of flowers, and delphiniums can be pollinated by insects other than bumblebees. Not so the Joshua tree, *Yucca brevifolia*, and the yucca moth, *Pronuba spinosa*. This particular symbiosis is so refined that, when new species of yuccas have appeared, so have new species of moths. The plants are totally dependent upon the moths for fertilization. When yucca flowers blossom, female yucca moths gather pollen into large balls, which they carry under their heads. When their supply of pollen balls is adequate, they climb partway up the pistils (female

Fig. VI-3. A yucca plant in bloom. Its flowers can be pollinated only by the yucca moth; in return, newborn yucca moths will have a ready supply of food in the form of yucca seeds.

portions) of the flowers, insert their ovipositors into the plants' ovaries, and lay their eggs. Then they ascend to the tops of the pistils and rub some of the pollen they have accumulated into the stigmas. Each female moth has gathered material for which she has no direct use and performed an act whose benefits she cannot perceive. Why? Certainly she has helped perpetuate the Joshua tree's existence. But she has also ensured the survival of her own species, for, as a result of her act of fertilization, the Joshua tree will produce seeds in the chamber where her eggs reside. Newborn moths will thrive on this immediately available food supply, although they will wisely (the wisdom obviously being built into their genes) not consume all the seeds.

Consider also the industrious mimosa girdler. This small beetle climbs out to the end of a branch of a mimosa tree. There it cuts a slit and lays within it some half dozen eggs. It then crawls back on the limb and labors for half a day, cutting a ring in the bark to the depth of the cambium. This final act offers no immediate gain—no food is extracted. But it does kill the branch's end, so that the next good storm knocks it to the ground. The beetle's eggs hatch in the fallen branch. There the hatchlings encounter the precise environment of dead wood in which they thrive. Moreover—and this is the surprising part—mimosa trees die after twenty-five to thirty years . . . unless they are vigorously pruned annually, in which case they can survive up to four times longer. Each, beetle and tree, derives a benefit from the other.

Not all the natural selective pressures toward cooperation and collaboration are interspecific. An example of intraspecific or group behavior is a school of fish. Since the range of visibility is short in sea water, it is not much easier for a predator to detect a group of fish than a single fish. Hence any one member's chances of being taken by a predator decrease inversely as the number of fish in the school increases (a piscine version of Russian roulette). Furthermore, a school, by dispersing in a sudden and evasive manner, can so confuse and distract a predator that it catches no prey. Schooling has advantages, too, when the fish are no longer prey but predators. Collectively they can herd *their* prey—that is, by spreading out, the school effectively casts a bigger "net." Clearly, individuals are enhancing their survival prospects by joining a larger assemblage that has some of the functional characteristics of a unit. The unit enhances the survival of particular genetic strains as well, because most members of a school are related. Therefore, the loss of an individual does not extinguish a particular genetic line.

Another example of what could be called a multiorganism—a collection of individual organisms behaving as a unit—is a bee swarm. Temperature regulation is extremely critical: excessive temperatures kill the swarm's inhabitants, while at too low temperatures bees cannot fly. An immobile swarm forfeits food and habitation sites to its competitors. By huddling more tightly together when ambient temperatures are cold and by spreading out when they are hot, bees collectively regulate the amount of insulation and ventilation in the interior of their swarm. Ventilation is effectively closed off in the tightly bunched swarm because bees in the outermost layer position themselves in a lattice of "shingles," the head of a bee below snuggling beneath the abdomen of its neighbor above. Simultaneously, heat generated within is trapped there. The layer of bees next to the outermost generates much of the swarm's heat by shivering. These techniques are so effective that the temperature at the core of a swarm can be held to within half a degree of 35°C. The fact that bees collectively can regulate the temperature of their environment does not imply they adhere to a plan originated and administered from outside. Individual members of the coordinated whole need not know that their activities work in such a way that the overall system produces conditions attuned to their survival. Rather, the coordinated system has evolved over eons of time by natural selection: bees that huddled together when cold produced more offspring than those that did not.

Fig. VI-4. Bees acting collectively to regulate the temperature inside their swarm. When external temperatures are low (left), bees pack closely together, and nearly every bee's head is beneath the abdomen of another. When the temperature is high (right), the bees spread out and orient their heads outward. (Courtesy Bernd Heinrich, University of Vermont.)

A prime example of the advantage of both cooperative, intraspecific behavior and interspecific symbioses is a termite colony. As with bees, a caste structure assigns specialized roles to the individuals in the colony. Together the termites are architects with prodigious capabilities. They construct elaborate homes, with exquisite control of temperature and humidity, by carefully stacking each other's fecal pellets. To the outside world, the entire colony appears to function as a single organism. Only close inspection reveals that it actually consists of individual organisms. An even closer inspection reveals that each individual termite is itself a complex ecosystem. In the hindgut of an Australian dry wood termite live several large (by the standards of single-celled creatures, at any rate) microorganisms (protists) called *Myxotricha paradoxa*. Without them the termite would starve to death. It could not digest the pulverized wood that constitutes its diet. Reciprocally, the protist enjoys a most accommodating environment, one buffered from the outside weather and continuously supplied with nutrients. But there is more! A *Myxotricha* cell is itself an ecosystem. For one thing, it must move about within the termite's gut to locate the richest source of nutrients. Alone, however, a *Myxotricha* cannot move about, so it lets other organisms propel it. Its motility is supplied by threadlike, undulating bacteria (spirochetes), some half million of which cling to its surface. Each spirochete lives in symbiosis with another kind of bacterium also attached to the surface of the *Myxotricha*. Yet a third kind lives inside the *Myxotricha*. These remarkable creatures are therefore both symbionts and hosts to at least three kinds of other symbionts.

The reality that single-celled organisms live inside other single-celled organisms—to the mutual benefit of both—makes symbiosis the probable mechanism in an evolutionary development that set the stage for rapid diversification of form among earth's biota. This development was the appearance of the nucleated cell. The earliest cells on earth were prokaryotes (pre-nuclear). Genetic material was dispersed throughout their interiors. Their only reproductive strategy was fission. Despite their undifferentiated interiors, some prokaryotes could trap sunlight via photosynthesis, others had threadlike bodies enabling them to wriggle about, and still others could oxidize or chemically decompose complex molecules, thereby releasing the energy locked within them. Eukaryotes, or nucleated cells, contain substructures (organelles) that carry out all these same functions: mitochondria for oxygen respiration, plastids for photosynthesis, and flagella or other motility organelles for propulsion. So similar are these organelles to individual prokaryotic cells that eukaryotes appear to have resulted from the incorporation of certain prokaryotes into others. Individual symbioses could have been blessed by natural selection. In other words, some prokaryotic cells were produced in greater numbers when they coexisted with some others than when they were not so associated, a situation that, over time, could have resulted in complete integration.

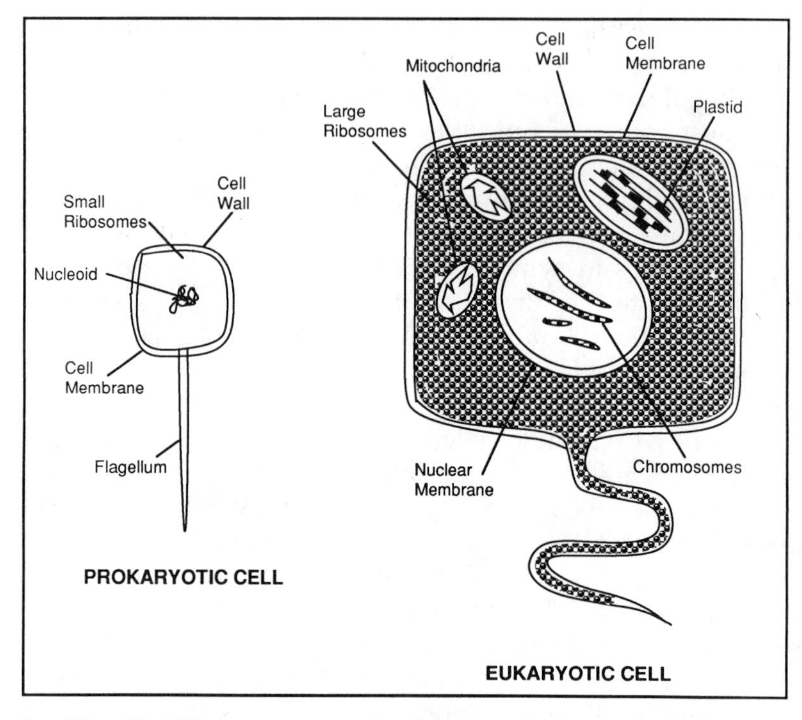

Fig. VI-5. The differing structures of prokaryotic and eukaryotic cells.

Eukaryotes, then, are examples of a cooperation and harmony so complete that a new organism is synthesized. That organism can no longer be subdivided into smaller living components; the latter are too interlocked into a greater whole to be able to exist in isolation. The eukaryotes' greatest contribution to life's continuity was the new reproductive strategy they introduced, the sexual union of two parents. An offspring that has received genetic material from two individuals is different from either parent. Diversity results from the many new genetic combinations that are forged. So symbioses that had advantages for all the cooperating (prokaryotic) partners eventually evolved into (eukaryotic) organisms capable of occupying new niches in the environment, capable, in fact, of creating new environments in which to reside. As a result, life's ability to convert incident energy into useful information was enhanced.

Each of the foregoing examples is fascinating in its own right, but all are presented to illustrate the strong tendency toward conjoined behavior among living specimens. The intent is to make integration of all life, as well as its (temporarily) nonliving setting, seem more feasible, even likely. After all, if individual

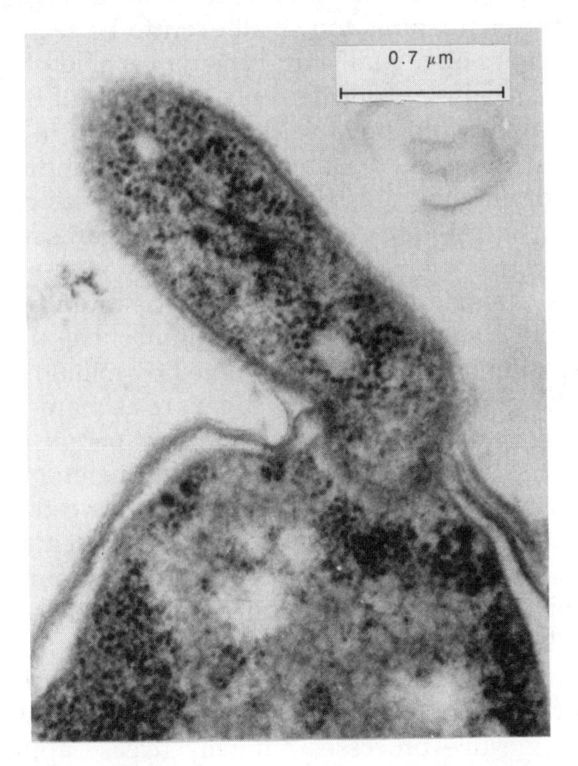

Fig. VI-6. Predatory bacteria. The predator, a Daptobacter *("gnawing") bacterium penetrates to the food sources inside its prey, a* Chromatium minus *bacterium. Is this a modern example of how in antiquity simple cells invaded each other until they co-evolved into internally complicated cells? (Photograph by Isabel Esteve and R. Guerrero.)*

organisms can actually merge into a larger living whole, is it inconceivable that this resulting whole can be a cog in a still larger unit, and so on until the unit incorporates the entire planet? This hypothesis was named after the Greek goddess, Gaia, by its founders, James Lovelock and Lynn Margulis. The actual "engineers" who carry out the functions of self-regulation and control that make the biosphere habitable—as, for example, bees make a swarm habitable—are microorganisms. We who are large in physique and powerful in mind may find it difficult to believe that our existence depends in large part upon the industry of creatures too small even to be seen. Can anything so small be so vital? Are bacteria capable of global engineering feats? If we consider only a handful of bacteria, the answer clearly is no. But the actual quantities in nature are staggeringly large. The product of the minute individual contributions of an enormous number of individuals can be arbitrarily large, and in the case of earth's microorganisms, it seems, it is.

Certainly a single termite's bacterial community, discussed earlier in the chapter, has only an infinitesimal effect on the global environment; but there are some fifty million times more termites than humans. In places there are as many termites in a square meter of land surface as there are people in a square mile (which is equal to 2.6 million square meters). Moreover, places where termites reside occupy about two-thirds of the earth's land surface. Taken together, these statements can be quantified to conclude that termites and ants account for about *one-third* of the total mass of animal matter in the Amazonian rain forest![1]

The summed activity of the huge population of bacteria residing in this enormous mass of termites is impressive. Their collective digestive processes annually release approximately fifty billion (5×10^{10}) tons of carbon dioxide to the atmosphere. To put this in perspective, it is about twice the global output

[1] Edward O. Wilson, 1985.

from fossil fuel combustion, that contribution by those other remakers of the global environment, we humans. The termites' bacterial output contributes more than a quarter of the planet's net primary carbon dioxide production. Less, although still staggering, amounts of other gases are also produced, among them 170 million tons of methane per year. One need not doubt that the bacteria within termites have the collective wherewithal to significantly alter the composition of atmospheric gases. And they represent just one sample from the total microbial census of the planet.

More evidence for the major role of single-celled organisms in the overall ecology of the planet can be gleaned from the oceans. On the average, a cubic centimeter of water (about a thimbleful) in the uppermost fifty meters of the ocean contains one and a half million organisms so small they are called picoplankton.[2] Like all plants, these use the sun's energy to fix the carbon from readily available carbon dioxide into organic compounds. Again, like most photosynthesizers, they also expire oxygen. As much as two-thirds of the total photosynthe-

[2] The name indicates that each organism's mass is about a picogram, or one-trillionth (10^{-12}) of a gram. Its diameter is several ten-millionths of a meter. The numbers provided illustrate the point that minute organisms in huge quantities can constitute a sizeable total biomass.

Every cubic centimeter of water contains $\sim 1.5 \times 10^{6}$ picoplankton, each of mass $\sim 10^{-12}$ g, or picoplankton density $\approx 1.5 \times 10^{-6}$ g/cm^3. The ocean covers about 70 percent of earth's surface, so

$$\text{ocean area} \sim 0.7 \times 4\pi\,(\text{radius})^2 \sim 0.7 \times 4\pi \times (6400\text{ km})^2 \sim 4 \times 10^{18}\text{ cm}^2.$$

The volume of ocean water to a depth of 50 m is approximately

$$\text{ocean volume} \sim \text{area} \times \text{depth} \sim 4 \times 10^{18}\text{ cm}^2 \times 5000\text{ cm} \sim 2 \times 10^{22}\text{ cm}^3.$$

Therefore the total mass of picoplankton is

$$\text{picoplankton biomass} \sim 1.5 \times 10^{-6}\,\frac{\text{g}}{\text{cm}^3} \times 2 \times 10^{22}\text{ cm}^3 \sim 3 \times 10^{16}\text{ g}.$$

For comparison, earth's 5×10^{9} humans, each of assumed mass ~ 50 kg, contribute

$$\text{human biomass} \sim 5 \times 10^{9} \times 50 \times 10^{3}\text{ g} \sim 2.5 \times 10^{14}\text{ g},$$

so

$$\text{picoplankton biomass} \sim 100 \times \text{human biomass}.$$

sizing activity of the oceans has been measured to occur within these organisms, so small as to be individually insignificant but so numerous as to be collectively dominant. The basic metabolism of the largest component of the biosphere, then, the hydrosphere, is driven by such nearly invisible organisms. The composition of the atmosphere, too, is largely determined by their industry. Picoplankton (e.g., cyanobacteria) began, about three billion years ago, to pour oxygen into an atmosphere that previously had contained very little of it. As they did so, they altered the global environment in a major way. Life adjusted to take advantage of the new opportunity created by transforming a reducing atmosphere to an oxidizing one: it evolved a capability to move out of the oceans and onto the land. Life could do so only when a particular molecular variant of oxygen called ozone (O_3) had built up in the atmosphere to such an extent that ultraviolet (UV) rays from the sun were filtered out. These rays, had they reached the earth's surface, would have been lethal. We stride on earth's surface only because previous generations of microbes made it possible for us to do so. Moreover, we are still strongly dependent upon the present generation of microbes to maintain and regulate the oxygen content of the air we breathe.

Production of various chemicals is only half of the equation. Consumption is equally important, otherwise the chemicals produced would accumulate without limit. The two, production and consumption, must operate as opposite sides of the same coin: what some discard, others must ingest. Furthermore, maintenance of a steady state requires regulation—feedback of current conditions so that the balance between productivity and consumption can be adjusted accordingly. Let us examine a few examples of regulated global engineering—engineering of Gaian proportions—to see how they might account for earth's favorable living conditions.

In a previous chapter we discussed the reciprocal relationship between plants and animals, photosynthesis in the one being rather precisely balanced by respiration in the other: carbon dioxide consumption and oxygen production by plants, oxygen consumption and carbon dioxide production by animals (see Figure V-6). What if a plant or an animal removes itself from this cycle by dying? In nearly all cases, the Gaian underworld of microorganisms takes over, decomposing the dead material into its original components. These latter are elemental nutrients again available for incorporation into living tissue and recirculation through the biosphere. Poet John Updike[3] has eulogized this process of rot, explaining more lucidly than any scientist that "all process is reprocessing:"

Der gute Herr Gott
said, "Let there be rot,"
and hence bacteria and fungi sprang
into existence to dissolve the knot
of carbohydrates photosynthesis
achieves in plants, in living plants.
Forget the parasitic smuts,
the rusts, the scabs, the blights, the wilts, the spots,
the mildews and aspergillosis—
the fungi gone amok,
attacking living tissue,
another instance, did Nature need another,
of predatory heartlessness.
Pure rot
is not
but benign; without it, how
would the forest digest its fallen timber,
the woodchuck corpse
vanish to leave behind a poem?
Dead matter else would hold the elements in thrall—
nitrogen, phosphorus, gallium
forever locked into the slot
where once they chemically triggered
the lion's eye, the lily's relaxing leaf.

[3] "Ode to Rot," from *Facing Nature,* by John Updike. Copyright © 1985 by John Updike. Reprinted by permission of Alfred A. Knopf, Inc.

All sparks dispersed
to that bad memory where the dream of life
fails of recall, let rot
proclaim its revolution:
the microscopic hyphae sink
the fangs of enzyme into the rosy peach
and turn its blush a yielding brown,
a mud of melting glucose:
once-staunch committees of chemicals now vote
to join the invading union,
the former monarch and constitution routed
by the riot of rhizoids,
the thalloid consensus.

The world, reshuffled, rolls to renewed fullness;
the oranges forgot
in the refrigerator "produce" drawer
turn green and oblate
and altogether other than edible,
yet loom as planets of bliss to the ants at the dump.
The banana peel tossed from the Volvo
blackens and rises as roadside chicory.
Bodies loathsome with their maggotry of ghosts resolve
to earth and air,
their fire spent, and water present
as a minister must be, to pronounce the words.
All process is reprocessing;
give thanks for gradual ceaseless rot
gnawing gross Creation fine while we sleep,
the lightning-forged organic conspiracy's
merciful counterplot.

The balance between consumption and production of vital chemicals can be disturbed; one way is burial of organic matter. If, before scavengers attack it, dead flesh or fiber is buried in sediments underlying a lake or an ocean, the carbon consti-

Fig. VI-7. "Nursing log" in Olympic National Park, Washington. Where a tree has fallen, microorganisms have decomposed it into nutrients that nourish the next generation of trees, neatly strung along the line of their fallen predecessor.

tuting the backbone of all its biomolecules is removed from the planet's metabolic processes. Every carbon atom so removed is one less to which atmospheric oxygen can be attached. Every removal therefore means less carbon dioxide and more oxygen in earth's atmosphere. Burial of as little as a tenth of a percent of the carbon once extant in living organisms suffices to account for the change in atmospheric composition to the present anomalous amounts of carbon dioxide (low) and oxygen (high).[4]

We have now introduced three interacting effects: photosynthesis, respiration, and burial of organic matter. How might these work together to maintain a global environment healthy to earthlife? Consider, on the one hand, a perturbation of the environment in the direction of inadequate oxygen supply. Then animals, fungi, some protists, and some bacteria suffer, among them the aerobic (oxygen-utilizing) scavengers. The burden of decomposition falls upon the community of anaerobes. It tries to break large organic molecules into smaller ones, but its mechanism, fermentation (chemical decomposition), is just not as efficient as oxidation (controlled burning). The inability of anaerobes to keep up with the supply of organic debris raining upon them means that more carbon-rich residues are buried. As a result, less oxygen is bound into atmospheric carbon dioxide and more is available to counteract the decrement there. On the other hand, if the atmospheric environmental perturbation had been in the opposite direction toward an excess of oxygen, the scavenging community would have proliferated. Little organic matter would have eluded it; burial of carbon would have diminished; more would have been available to remove free oxygen from the atmosphere by forming carbon dioxide. Simultaneously the increased release of methane to the environment resulting from the decomposition of sugars, carbohydrates, and proteins would have engaged more of the oxygen there, also helping to reduce its excess.

This is but one (oversimplified) example of life regulating the

[4] Anomalous with respect to atmospheres of neighboring planets:

	CO_2 Content	O_2 Content
Venus	97%	$\leqslant 1\%$
Earth	$\leqslant 1\%$	21%
Mars	95%	$\leqslant 1\%$

chemical composition of the global environment—and doing so to its own advantage. The significance of this example is not just that nutrients for living organisms are recycled, but that they are reentered into a fluid medium—the atmosphere or the hydrosphere—that can transport them to locations where biota can use them. A second example of organic manipulation of the biosphere explains its temperature regulation. Such gases as carbon dioxide, ammonia, and water vapor act upon energy flows like the rolled-up windows of an automobile: visible sunlight readily passes through them, but infrared radiation emitted by the objects the sunlight warms does not; it is absorbed. As a result, heat is trapped in earth's atmosphere by a layer of carbon dioxide existing there. The atmosphere contained more carbon dioxide early in its history than it does now. In addition, it received less solar energy then than now. Yet earth's surface temperature has scarcely varied. Could the greater blanket of carbon dioxide on primitive earth have trapped more of the weaker solar energy available? And could the amount of carbon dioxide have been gradually reduced *by living organisms* exactly at the rate required always to balance its insulating qualities against the growing energy input? If so, life collectively has created, as opposed to passively accepted, the conditions under which it prospers.

The picoplankton that were discussed earlier may also play a role in maintaining a stable thermal environment, for they produce a chemical compound called dimethylsulfide (DMS) that diffuses its way from the seawater in which it is released to the atmosphere. Once there, it reacts chemically to form particles upon which water vapor can condense into liquid droplets. The droplets constitute clouds, and when enough form in the atmosphere the sun's energy is reflected back into space before reaching the surface, which consequently cools. Now if, as suspected, picoplankton produce less DMS when temperatures drop, a correcting mechanism is introduced. Follow the sequence: temperature at the surface falls, picoplankton produce less DMS, fewer clouds form in the atmosphere, more sunlight strikes the surface, and temperatures there rise. If, on the other hand, surface temperatures climb, enhanced picoplankton production of DMS acts to reduce them.

A prodigious engineering feat by organisms functioning to maintain a habitable environment *may* explain the constancy of the oceans' salinity. The source of salt in the sea seems obvious: rain falling on rocks and soil washes some of their minerals into the seas. In fact, it washes about 540 million tons of salt into the ocean every year. At that rate, all the salt in today's hydrosphere would have been deposited there in just eighty million years, actually less, because salt also enters with the magma pushed from the earth's interior through the ocean floor. But geological evidence supports the thesis that the salt content of the ocean has been constant for billions of years. The problem is not how to provide salt to the oceans; it is how to remove it. One known mechanism occurs at lagoons, shallow bays, and other isolated arms of the sea. If evaporation occurs faster than input from rivers, springs, and rain—as, for example, at the Persian Gulf—huge beds of salt are laid down. Eventually these will be buried by natural geologic processes, as the present giant salt fields under the midwestern United States attest. Are there places where life is arranging massive deposition of salts for removal from the ocean? The Great Barrier Reef of Australia may be one instance of an evaporative pond under construction by the cooperative behavior of the many kinds of living organisms in a coral community. The timescale of construction appears to ensure completion at approximately the epoch when some salt will have to be removed from the hydrosphere to prevent an increase of its salinity.

The proposition that all life can work together to support and sustain itself raises the question of the boundary of the living. Specifically, defining *life* as *a system able to decrease its internal entropy at the expense of a flow of free energy from its external environment* implies a distinction between internal and external. But the interdependencies of all organisms and the nonstop

Fig. VI-8. Great Barrier Reef of Australia off Heron Island. Is the coral biocommunity constructing an evaporative pond in which salt will eventually be removed from the ocean?

recycling of raw materials and energy through the living and the nonliving portions of the earth mean that the entropy changes are coordinated over the entire planet. A cosmic perspective reveals a *visible* boundary to the superorganism at the top of the atmosphere, but this evidence is deceptive. *Invisible* influences extend much farther than the top of the atmosphere. Earth's magnetic field, for instance, reaches well into interplanetary space. Life benefits as a result, for the magnetic field deflects the flow of energetic cosmic particles capable of killing organisms they strike. A record of the earth's magnetism is recorded in rocks, especially those extruded at spreading seafloor ridges. Surprisingly, the record indicates periodic reversals of the north and south magnetic poles. Even more surprising is the recent evidence that reversals of the magnetic field's polarity may coincide with the onset of epochs of unusual cold and with sudden increases in the amount of fragmentary matter of extraterrestrial origin deposited on earth's surface. Here are three global phenomena—field reversals, extraterrestrial matter bombardment, and cold eras—that *may* have cause-and-effect relationships (the theories are admittedly speculative). A heavy influx of matter could, for example,

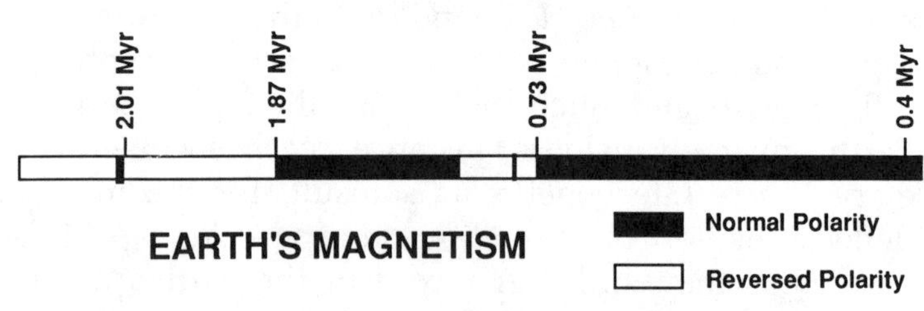

Fig. VI-9. Lake sediments. Scientists can determine the magnetism and the quantity of organic matter in each layer. A decrease in organic matter signals a drop in temperature. Note that each sharp temperature drop coincided with a reversal of earth's magnetic polarity. Data from the dry lake bed constituting the Karewa Plateau in the Vale of Kashmir. (Krishnamurthy, R. V., Bhattacharya, S. K., and Kusumgar, Sheela. 1986. Nature *323: 150–152.)*

darken the skies, thereby blocking sunlight from reaching the surface. The resultant cooling could add more ice to the polar caps at the expense of liquid sea water. If sea level falls and polar ice mass rises, the planet has effectively undergone a redistribution of mass, one possibly significant enough to alter the earth's rotation. But the dynamo at earth's center generating its magnetic field might then undergo a temporary readjustment, out of which might come a reversed magnetic polarity. The whole point is that cosmic events can trigger global changes that ripple through the biosphere, producing major changes within it. Each cosmic trauma spells doom for some organisms and creates opportunity for others. The biosphere, it appears, is linked to matter and energy organized well beyond earth's boundaries.

We have already mentioned that the energy driving the biosphere's metabolism derives to a very large extent from the sun, which therefore has to be considered part of the overall living system. The linkage is so critical that even small changes in solar output, which occur when a tiny fraction of its surface is darkened by spots, produce significant effects on earth's climate, hence on its biota. The energy received from the sun can also vary as a result of slight cyclical changes in the earth's orbit. The critical orbital parameters are the tilt of the earth's axis to the plane of its orbit, which varies with a periodicity of 41,000 years; the direction in which the axis points, which has a period of 23,000 years; and the eccentricity (departure from circularity) of the orbit, with a period of 100,000 years. The glacial record

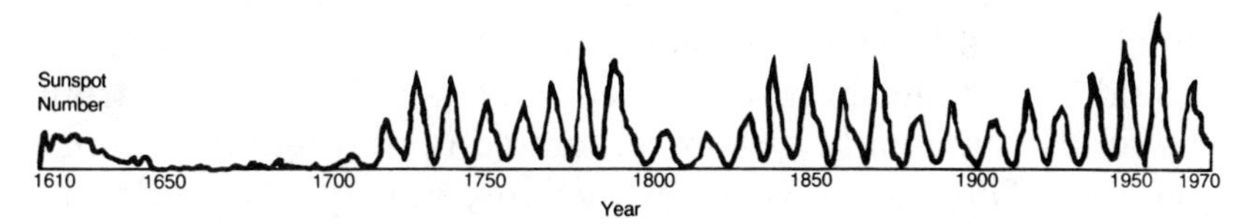

FIG. VI-10. The numbers of spots on the face of the sun over the past three centuries. Almost no spots appeared between 1645 and 1715, during the so-called Maunder minimum, the same dates encompassing the Little Ice Age. Coincidence? (NASA.)

stored on earth reveals advances and retreats of ice in synchrony with these orbital periodicities, as predicted by Milutin Milankovitch. Milankovitch's theory establishes that the principal causes of the succession of ice ages over the past 800,000 years, events that certainly had dramatic effects on the evolution of the biosphere, are subtle but predictable patterns in earth's orbit resulting from the mechanics of the entire solar system.

The sun is not the only object in the solar system affecting life's destiny. Asteroids swirl around it in an orbital belt between Mars and Jupiter. They are, in essence, rubble that perversely avoided agglomeration into a planet during the solar system's formative years. Many thousands exist, from pebbles to kilometer-size rocks. Frequently they pepper planets and moons, leaving craters of various sizes in their wakes. As the earth and this rock field play "solar system billiards," a collision with a large asteroid can be expected about once every hundred million years or so. A collision with a body, say, 10 kilometers in diameter is not a footnote in biological history, but a headline. Strong evidence suggests that just such a collision occurred 65 million years ago. The evidence resides in the strange chemical composition of a worldwide layer of sedimentary rock deposited at that epochal date. The sediment is enriched with iridium, an element not abundantly present on earth but much more so in asteroids. Presumably one shattered upon impact with the Earth, and its fragments, together with the soil ejected from the crater it carved, spread around the globe. Enough matter could have polluted the atmosphere to diminish significantly the amount of sunlight reaching earth's surface. A colder climate would have resulted. In addition, photosynthesis would have been sharply reduced, and its disappearance from the food chain would have reverberated up through the energy pyramid. The result? Major extinctions, exactly as recorded in the paleontological record of 65 million years ago. The most sorely missed species eliminated by this astronomical catastrophe are the dinosaurs. Of course, their elimination, coupled

Fig. VI-11. Meteor Crater, Arizona. Some of the material excavated from this crater, as well as that of the body that caused it, could have darkened the skies and cooled earth's surface dramatically.

with that of the many reptilian relatives that accompanied them to oblivion, was not all bad news. Mammals, in fact, saw the newly vacant niche as a welcome opportunity. Their rapid radiation led eventually to creatures capable of reading this book.

Recent evidence suggests that similar events occurred repeatedly throughout earth's history. Both the record of cratering and the fossil record of extinctions hint at periodicities in the range of 26 to 30 million years; furthermore, the epochs of major cratering and mass extinctions coincide. Recurring phenomena, particularly those with such long timescales, suggest cosmic rhythms. Where else but the cosmos do events repeat so faithfully? Might the earth be bombarded at regular intervals by swarms of comets and asteroids? Yes, hypothesize some, if

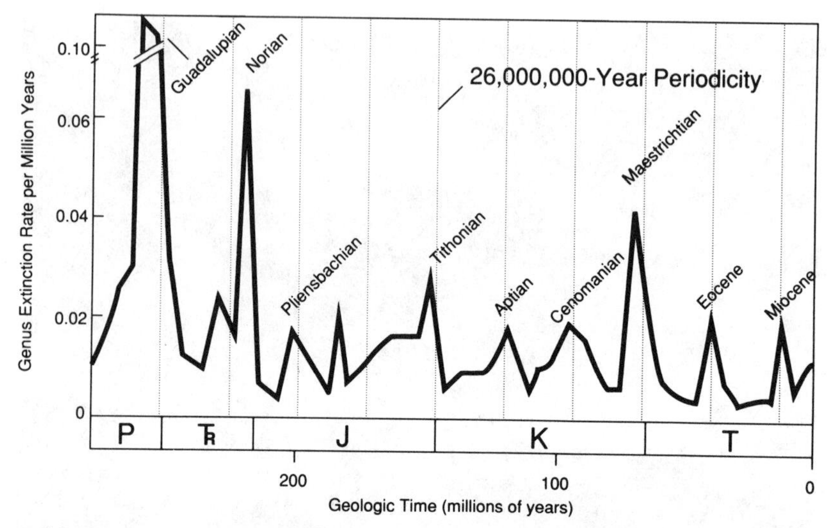

Fig. VI-12. Extinction rate over past 250 million years. The light vertical lines recur every 26 million years. Extinctions seem to peak synchronously with this interval. (Courtesy J. John Sepkoski, Jr. 1988. Los Alamos Science ***16:*** *36–49.)*

a small companion star to the sun repeatedly swings near by in its orbit, or if the sun regularly bobs up and down through the midplane of the galaxy, encountering large galactic clouds when it does so. Either scenario exposes the solar system to a gravitational perturbation that jars asteroids from their orbital belt and sends them plummeting into the inner solar system. If the history of life's evolution is regularly punctuated by these occurrences, our links to the cosmos are almost tangible.

The boundary enclosing the living superorganism of which we are a part can clearly be stretched across the entire solar system, especially when we consider its time dimension as well as its spatial ones. But objects and events beyond the solar system are also registered in our bio-history. Consider the death of a star in the eruption of a supernova. This devastating event sprays potent radiation (X- and gamma-rays, for example) and penetrating particles into the space around it. Such showers from these stellar remnants would be lethal to some life and would increase mutations among survivors. The effect on the biosphere cannot be ignored, since supernovae are thought to erupt within 30 light-years of the solar system every hundred million years or so.

Clearly we, the collective membership of a superorganism, are products of an environmental setting stretching light-years in space and "billenia" in time. Our integration into a unit has been difficult to recognize because, until very recently, we were lost in the inner workings of it. Our plight resembled the way a neuron in one's brain or a muscle cell in one's big toe, preoccupied as it is with carrying out its individual function, is unaware of its participation in a larger whole. But we humans, both physically in our spacecraft and mentally in our imaginations, have stepped outside the boundary within which the entropy reduction to which we contribute takes place. As a result of having an outsider's vision and an insider's role, we can make the energy-for-order conversion more efficient . . . or we can destroy it altogether. We can even spread order to other places in the solar system and then to the galaxy. But we demand a lot of the bacteria crucial to maintaining Gaia's homeostasis if we continue to manipulate changes of the magnitude and on the timescales of our ongoing rampant development. The fact that Gaia has restored balance following every previous perturbation is comforting, but illusory in the sense that we have never been here before. Our arsenals, chemicals, earth-shaping equipment, and population sprawl may occasion changes no larger than previous catastrophes, but the pace with which they are advancing may be faster than the natural time constants of the system called upon to respond to them. A human organism, for an analogy, can keep up with the tissue deterioration of the aging process, but not always with the instantaneous deterioration following a car wreck. Until certain knowledge of our impact replaces present ignorance, our best strategy is to tread softly.

For Further Reading

Alvarez, Luis W. 1987 July. Mass extinctions caused by large bolide impacts. *Physics Today* 40(7): 24–33.

Covey, Curt. 1984 February. The earth's orbit and the ice ages. *Scientific American* 250(2): 58–66.

Gould, Stephen Jay. 1984 August. The cosmic dance of Siva. *Natural History* 93(8): 14–19.

Heinrich, Bernd. 1981 June. The regulation of temperature in the honeybee swarm. *Scientific American* 244(6): 146–160.

Kerr, Richard A. 1980. Asteroid theory of extinctions strengthened. *Science* 210: 514–517.

Kerr, Richard A. 1983. Orbital variation-ice age link strengthened. *Science* 219: 272–274.

Kerr, Richard A. 1984. Periodic impacts and extinctions reported. *Science* 223: 1277–1279.

Kerr, Richard A. 1987. Milankovitch climate cycles through the ages. *Science* 235: 973–974.

Krebs, John R. 1979. Coevolution of bees and flowers. *Nature* 278: 689.

Larson, Peggy, and Larson, Lane. 1977. *The Deserts of the Southwest*, pp. 105–106. San Francisco: Sierra Club Books.

Loon, Owen B., and Olson, Steve. 1985 October. The warm earth. *The Sciences* 26(4): 50–57.

Lovelock, J. E. 1979. *Gaia*. Oxford: Oxford University Press.

Margulis, Lynn. 1971 August. Symbiosis and evolution. *Scientific American* 225(2): 48–57.

Margulis, Lynn, and Lovelock, James E. 1981. Atmospheres and evolution. In *Life in the Universe*, ed. John Billingham, pp. 79–100. Cambridge, Mass.: The MIT Press.

Monastersky, Richard. 1987 December 5. The plankton-climate connection. *Science News* 132: 362–365.

Partridge, Brian L. 1982 June. The structure and function of fish schools. *Scientific American* 246(6): 114–133.

Raup, David M. 1986. *The Nemesis Affair*. New York: W. W. Norton & Company.

Sagan, Dorion, and Margulis, Lynn. 1986. *Microcosmos*. New York: Summit Books, pp. 115–136.

Sagan, Dorion, and Margulis, Lynn. 1987. Bacterial bedfellows. *Natural History* 96(3): 26–33.
Schwarzschild, Bertram. 1987 February. Do asteroid impacts trigger geomagnetic reversals? *Physics Today* 40(2): 17–20.
Simberloff, Daniel. 1984 July-August. The great god of competition. *The Sciences* 24(4): 16–22.
Thomas, Lewis. 1981 July. Debating the unknowable. *The Atlantic Monthly* 248(1): 49–52.
Thomas, Lewis. 1984 August. An argument for cooperation. *Discover* 5(8): 66–69.
Wilson, Edward O. 1985. The sociogenesis of insect colonies. *Science* 228: 1489–1495.
Weiner, Jonathan. 1986 July-August. In Gaia's garden. *The Sciences* 26(4): 2–5.
Zimmermann, P. R., Greenberg, J. P., Wandiga, S. O., and Crutzen, P. J. 1982. Termites: a potentially large source of atmospheric methane, carbon dioxide, and molecular hydrogen. *Science* 218: 272–274.

SEVEN

Natural Selection for Self-Observation

The most complicated material object in the world as we know it is the human brain. The brain is not a clockwork machine. It is not a digital computer. We are part of that complex web of natural selection which has itself evolved a selection machinery called our brain. In each one of us lies the richness of a second evolutionary path during a lifetime: it unites culture with a marvelous tissue in which the hope of our survival lies.

Dr. Gerald Edelman

LIFE IS CHARACTERIZED by the spread of order, albeit at the expense of a greater disorder exported elsewhere. The increase of order over time has been possible because life has steadily improved the efficiency with which it exploits the flow of energy on which it thrives. Two strategies have evolved for maximizing energy exploitation: (1) maximization of the numbers of individuals within each species, and (2) diversification of species, so that one organism can extract order from what another has rejected as chaos. Both strategies are possible only if life persists for a long time. Its thread of continuity, once initiated, can never be broken, or succeeding life forms will have no predecessors upon which to build.

The dimension of time is woven into the very definition of life. Life has been continuous because organisms strive to reproduce themselves. And to close the circle of the argument, organisms viable today are those that have reproduced themselves, because natural selection has weeded out any nonreproductive variants after a single generation. To illustrate the great efforts organisms expend to reproduce, to perpetuate themselves by parenting offspring, consider the bowerbirds of New Guinea and Australia. Reproduction is almost their central activity. Male bowerbirds invest enormous time and energy erecting elaborate architectural structures specifically to attract mates and to signal dominance to competing suitors. These structures include walled avenues, large woven huts, tall maypoles, carefully manicured lawns, and moss platforms with parapets. Often they are oriented in an approximate north-south direction, perhaps so that male and female can face each other in the early morning with neither looking into the sun. The males enhance the allure of these love parlors by decorating

*FIG. VII-1. Male (foreground) and female (background) bowerbirds (*Amblyornis flavifrons*) and their bower. The male has built the bower by piling up sticks about the fern, constructing a platform of moss beneath it, and stacking blue, yellow, and green mounds of fruit. (Jared M. Diamond. 1982.* Science ***216:*** *431–434. Copyright © 1982 by the AAAS.)*

them with brightly colored objects: shells, mushrooms, fruits, flowers (changed daily), bones, and, recently, coins, film cartons, buttons, bottletops, and other human paraphernalia; or by tastefully painting them using a tool dipped in crushed fruit, charcoal, or blue laundry powder. The colors are not arbitrarily chosen. In most cases, they either match or complement prominent colors in the birds' own plumage.

A male bowerbird's seductive strategy is equally impressive. He is an accomplished vocal mimic, capable of many sounds and rhythms, which he orchestrates into a variety of songs.

When a female is attracted to his vicinity, he presents an erect crest of spectacular plumage.[1] With exquisite aesthetic taste, he also displays a bright, colorful object, held in his bill and advanced in such a way that the female sees it always against the backdrop of his own brilliant plumage. The object has been carefully chosen to match the bird's plumage. In addition to perfecting his own appeal, a male devotes some time to lessening that of his competitors. Some of his own ornaments have been stolen from rivals. Even vandalism against another's bower is practiced.

Obviously, reproduction to a bowerbird is not a minor diversion from a central preoccupation with staying alive. It is a major part of its life cycle. Selfish concentration on its own existence is sacrificed to ensuring the existence of successors. And the same is true of virtually every other case in the natural world. Investigators seeking a way to exterminate tsetse flies, carriers of sleeping sickness, discovered that female flies release a chemical sex attractant so potent that a male will carry an impregnated decoy sixty times heavier than himself until he dies from exhaustion! Some animals seem to live only to reproduce. Spadefoot toads, for instance, lie buried in desert mud for a good ten months of every year until a summer rain stimulates in them an intense burst of mating behavior. Their offspring have to progress from egg to tadpole to toad before the puddle in which they are germinated dries up, an interval that may last only a week or two in the desert. Those fortunate enough to survive these accelerated stages of fertilization, birth, and growth must burrow to safety themselves as the summer rains depart. There they lie dormant for the majority of another year. And spadefoot toads are not the extreme of sexual behavior! Consider Atlas moths. They emerge from the chrysalis without mouths or digestive systems; they immediately begin starving to death. They live only long enough to mate and lay

[1] Not all species of bowerbirds have crests. In general, the less spectacular the plumage, the more elaborate the bower.

eggs. An entire phase of their life cycle is devoted solely to reproduction. These moths exemplify to perfection the thesis of sociobiologists that "an organism (usually a chicken) is just an egg's way of making another egg."

Why? What immediate benefit accrues to the reproducers? Why should the present generation care about posterity? After all, what has it done for us lately? It seems as if we—humans, bowerbirds, tsetse flies, spadefoot toads, Atlas moths, and all the rest—are caught in a flux, a lifestream, that propels us onward. We are transient vehicles arranged by the information sources we contain because they find us convenient for preserving and for passing that information into the future. Metaphorically we run a leg of an ever-branching relay race, passing on a baton of order that will both shape the next runner in the relay and present to him an information pool to which he can add before handing it to his successor. The race is peculiar because the baton, not the runners, drives it onward. Life's spread of order resembles a cancer: once started, its growth cannot be stopped.

All life is interdependent not only with all other concurrent life, but also with all preceding life. Just as no individual organism can survive without access to the byproducts of the activities of some others, no individual life cycle, birth to death, constitutes a meaningful unit. Today's life, a collectivity, is the integration of all contributions, historical as well as current.

We are immediately confronted with two fundamental questions. How is the information (or order) that life represents reproduced? And how has that information been able to increase with time? The first question asks how parents produce like offspring, and the second asks how variants come into being. The answers to both lie in the structure of the DNA molecule, a molecule readers have encountered in previous chapters. If the immensely long coded message carried by the specific sequence of the four types of nucleotide in a particular DNA molecule were spelled out linearly, it would, like the slim

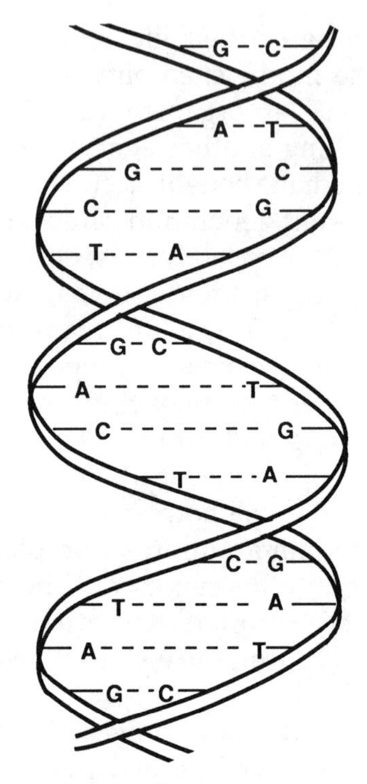

Fig. VII-2. The helical structure of DNA. The sequence of nucleotides on one strand is dictated to be the complement of the sequence on the other strand.

stalk of a very tall flower, be subject to fragmentation. If instead it resembled a vine of ivy—one strand intertwined with another—its integrity could be preserved. And this latter configuration is precisely what it has selected. The supporting structure with which either strand of DNA is intertwined contains precisely the same information as the other but written in a complementary language. The two strands are joined together in a helical coil such that an A nucleotide in one connects with (and only with) a T in the other, likewise a C with a G[2].

[2] A, T, C, and G for Adenine, Thymine, Cytosine, and Guanine, respectively (see Figure V–4).

These cross connections, as well as the lengthwise ones that bend each helix, are consequences of the distributions of electric charges that surround each geometrically precise molecular unit.

The electrical attraction each nucleotide exerts on its complement explains how DNA replicates itself. If the intertwined helices unravel into their component strands in a medium rich with the four kinds of nucleotides, each strand re-assembles its complement link by link. An unattached T is naturally drawn to an A on one strand of the unraveled helix, while an unattached A bonds with the T that formerly connected with the first strand's A. The same for C's and G's. Eventually there will be two strands exactly like the first. Note that the bonding properties of DNA must be special: on the one hand, lengthwise bonds must be very strong to preserve the integrity of the sequence; on the other hand, inter-helical bonds must be weak enough to permit unraveling but strong enough to attract complements. Upon this happy interplay of electrical forces does the continuity of life rely.

Having seen how the information content of life is duplicated in a package that can be propagated forward in time, we can now address how that information continually increases, how new varieties of organisms are created. One way, at least for multicellular organisms, is via sexual mixing of the genetic contributions from the organism's parents. Every cell of an organism contains two sets of DNA, one from each parent, every cell, that is, except the sex cells. To avoid the impossible situation of four strands of DNA in one's children, then eight in the grandchildren, and so on, both the sperm and the egg must reduce their DNA by half. This task is accomplished in the testes and ovaries, respectively, where approximately equal portions of each set of the parental DNA are assembled lengthwise into a single strand. This is the strand that meets another, similarly produced, in the sex cell of a mate to produce a fertilized egg, which has, again, two sets of DNA. Note how many sources have contributed information to the egg that will be a

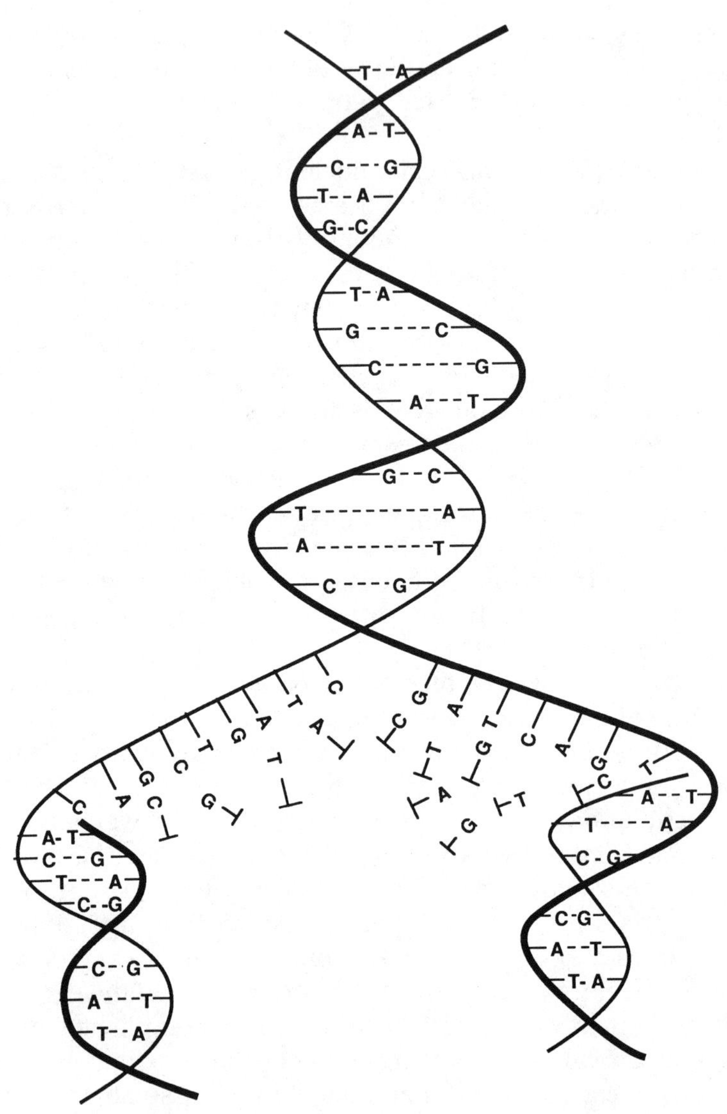

Fig. VII-3. If a DNA molecule unravels, each molecular site of each strand attracts the appropriate complementary nucleotide. The final result is two helices identical to the original one.

member of the next generation: each parent donates part of his or her father's and part of his or her mother's DNA. Follow this trend yourself for just a few generations to see how change accumulates. Sexual mixing of genetic information introduces variety to the gene pool very quickly.

Change is introduced in another way. Occasionally, radiation or deleterious chemicals, as two specific examples from among many, damage a nucleotide link in the DNA of an organism. The link may be transformed to another nucleotide, altered into a form that is not one of the four standard ones, or removed from the chain altogether. Almost always the mutation is detrimental. The DNA before alteration was, after all, the successful product of billions of years of testing for survival value. It is unlikely that random fiddling would lead to a more fit survivor, just as it is unlikely that random rearrangements of the letters on this page would produce a more readable manuscript. But the unlikely does happen. Some mutations produce improvements, just as, to continue the analogy, judicious editing improves manuscripts. If the mutations occur in the sex cells, they acquire the quasi-permanence that characterizes sexual reproduction. So a second scheme for introducing variety exists. There are many more, but explaining them all is the task of a biology text, not of a book on the living universe.

One result of the increase in diversity is to guarantee life's continuity by creating sufficient options that some can survive any change, gradual or abrupt. One of the most recent changes is also one of those most essential in perpetuating life's order, namely, the introduction of intelligence. If life's only transfer of information were by direct passage of genes between generations, the exchanged information content might never exceed some maximum. How long, for instance, can a molecular code be strung together before it loses its structural integrity? Furthermore, information transferred only genetically lacks the flexibility necessary to respond quickly. No genetically based change can occur in less than several generations, and for large

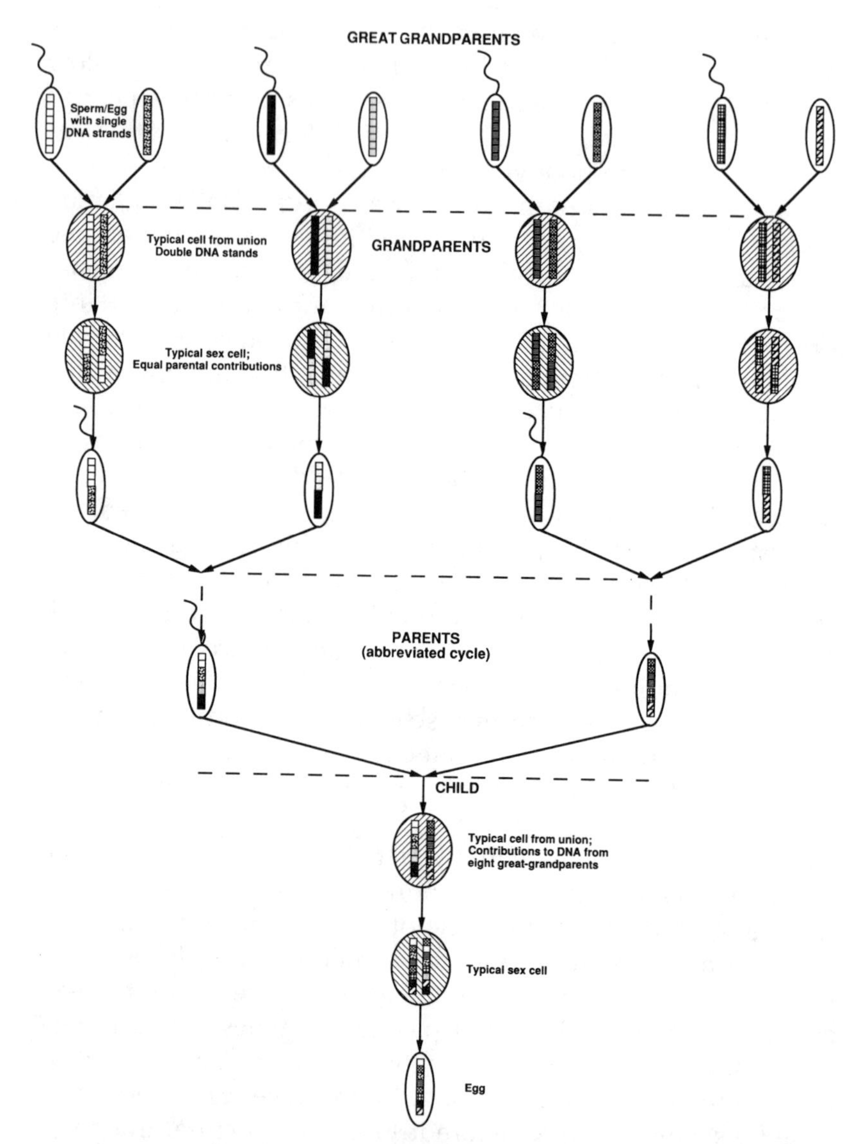

FIG. VII-4. Sexual reproduction contributes variety to the gene pool. Two parents, for example, produce offspring whose genetic endowment includes contributions from four grandparents and eight great-grandparents.

organisms the time between successive generations is long. They therefore need other means for rapid adaptation, or life will be limited in size, and some niches left unoccupied.

Intelligence became the means permitting rapid adaptation. Its application in practice is evident in many situations, but we choose to consider again the bowerbirds to illustrate it. Are their courtship rituals instinctive, preprogrammed activities, or are they taught by one generation and learned by the next? The evidence suggests that many of the skills are learned. For instance, juvenile males spend about two years constructing rudimentary bowers. Occasionally their clumsy efforts are supervised by mature adult males. The youth do not paint their bowers, nor is their choice of decorations as aesthetically exquisite as it is among their elders, but they are conducting trial-and-error experiments. When adult males conduct their bower activities with greatest intensity, the most keenly observant onlookers are the young males. All this suggests that bowerbirds learn by example, a sure sign of intelligence. To some extent this learning that intelligence permits is a luxury. The bowerbirds, for example, need ample leisure time to be able to court mates so assiduously. In this their environment assists them, for it has abundant supplies of concentrated fruit crops high in nutritional value. A quick meal can therefore fuel a long period of fasting.

Stepping back to view intelligence from a general perspective, we repeatedly find, as with the fruit resources for bowerbirds, that the environment is a strong factor in its rate of development. Since complicated responses are more essential to inhabitants of complex environments, environmental diversity should foster higher intelligence. All evidence indicates that it does. Marine organisms, occupants of rather homogeneous surroundings, lag behind their terrestrial counterparts in encephalization.[3] In fact, the most highly encephalized creatures

[3] Encephalization is a measure of the fraction of a body's mass devoted to brains.

in the seas are those mammals who reinvaded them from the land, e.g., the whales and dolphins. Further evidence is at hand in the comparison between fruit- and insect-eating bats. The frugivores, who must search for packets of energy distributed erratically in space and time, have more brain tissue than the insectivores, whose energy sources are ubiquitous.

One factor in determining an environment's diversity is its dimensional complexity, the degree of freedom available to an organism for movement. Flying insects and birds, able to move freely in all three dimensions of space, have greater neurophysiologic capacity than more sedentary animals. Likewise, free swimming sharks are more encephalized than the creatures who are confined to the two-dimensional world of the sea floor. It is the active creatures who are able to exploit fully the dimensional complexity of any environment. They are the ones who eat high on the food chain and forage extensively. For them encephalization permits sensory integration, flexible behavior, and quick response. So advantageous are these features that the fractions of their overall weights devoted to their brains are larger than in creatures whose lifestyles are relatively static, e.g., the ones whose defenses are spines, armor, or poison.

All this encephalization is expensive. Large brains present large energy demands. In humans, a fifth of their energy consumption supports the operation of their brains. This fraction by itself is as great as the energy required for *all* the metabolic functions of an active lizard. Clearly, a lizard's energy budget prohibits a large brain. Only creatures with high metabolic rates can afford large brains, which explains why birds and mammals have the highest levels of encephalization. Not all the cost of high intelligence is measurable in energy terms. There is also a price paid in time, specifically for learning. Humans, boasting the largest encephalization parameter, dramatize this through the immense investment of parental upbringing required to produce a smoothly functioning organism. Nothing in the world is more poorly equipped to survive in it than a newborn human. But nothing has as great a potential to survive any-

Fig. VII-5. The margay (top), a desert wildcat, needs large reserves of brain power to make the instantaneous decisions essential to a predator. The desert tortoise (bottom), safe from attack and consumer of stationary food sources, needs less.

where in that world, or even beyond it, as the same human nurtured on the knowledge his ancestors have accumulated.

The ability of intelligent organisms to exist in less than optimal biological settings illustrates how intelligence increases the efficiency with which life converts energy to order. Constraints that limit the distribution of earth's less encephalized organisms are "outwitted" by its modern, intelligent inhabitants. The most encephalized residents of the poles, deserts, and oceans—regions of great environmental homogeneity—are those creatures that arose in more biologically diverse environments but emigrated from them to exploit relatively virgin niches. The extreme of emigration has occurred in the immediate past: we humans have ventured into space. This act demonstrates not just the convenience but also the necessity of in-

FIG. VII-6. Intelligence permits organisms to function in environments for which they are not biologically equipped. (NASA.)

telligence for the maintenance of life; for at last we can glimpse the possibility that the ultimate fate of earth's biota need not be linked to the sun. The sun's energy is today the source upon which nearly all life feeds. But it cannot be the source forever; its lifetime is finite. Knowing this, intelligent life can begin to make preparations. Among possible survival schemes needed within the next five billion years might be the mobility to travel to other energy sources or the ability to "astro-engineer" new ones in our neighborhood.

Are such feats unrealizable fantasies of speculative science fiction? Who can say with certainty? But the rate of blossoming of intelligence in our past, extrapolated into the future at the same pace, offers almost limitless potential. In a mere 900,000 years, human brains may triple in volume. To illustrate what that might mean, note that the doubling that occurred in the past two million years fostered the development of languages, agriculture, cities, and tools so advanced their users can probe all of nature. Human slowness afoot has been not only com-

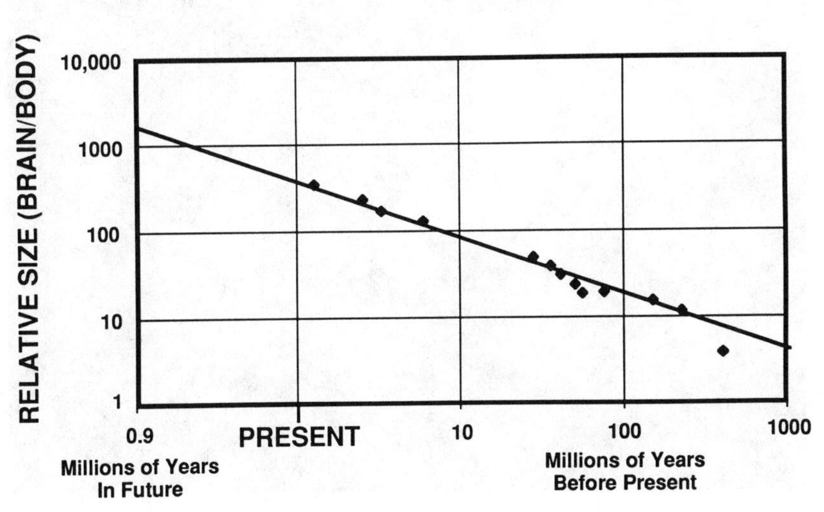

FIG. VII-7. Extrapolation of brain size from the past into the future. In 900,000 years brain size may be triple present size. (Data: Russell, 1981.)

pensated but totally eliminated by the wheels we have put beneath us and the wings above us. Our limited vision is extended outward by the telescope and inward by the microscope. Accelerators unveil the fundaments of matter, simultaneously probing the earliest moments of cosmic history. Our feeble voices are amplified and transmitted at the speed of light to the vastness of interstellar space. Our ephemeral existence leaves a permanent legacy in the written word, the photograph, and the recording. All these extensions of our being, tools of our own making, are detachable. Our "eyes" have seen the backside of the moon and the nuclei of our cells. Our "ears" have heard thunderstorms on Jupiter and the heartbeat of a human embryo. Our "mental clocks" have revealed the very beginning of the universe and have forecast several billenia hence. And these are just past achievements, certain to be overwhelmed by those in our future.

Fig. VII-8. Voyager 1 Spacecraft has extended human senses throughout the solar system. (NASA/JPL.)

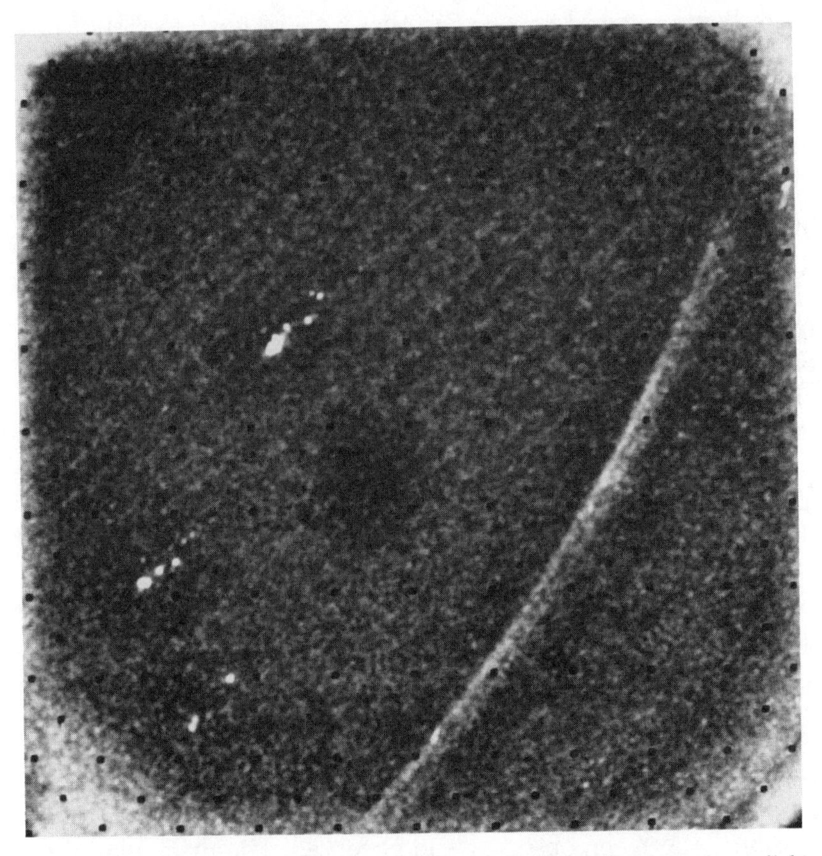

Fig. VII-9. Humans have "heard" thunderstorms on Jupiter (bright spots are lightning) and have "seen" its aurora (long arc). Data from Voyager Spacecraft (NASA/JPL.)

With intelligence, nature has created a way of observing itself. We who are parts of nature and also blessed with the intelligence to observe it constitute a self-referential paradox. Are we not akin to the artist in his studio who paints a picture of that studio, containing, of course, himself painting a picture of the studio? The picture in the picture likewise includes the same artist painting a picture of the same studio, and so on *ad infin-*

itum. Being integral to a system that we also wish to view from a detached perspective presents a profound conundrum. Nevertheless, it seems as if we humans can reconstruct our past and can help determine our future (or whether we care to have one). We can also map out our spatial surroundings. Knowledge of where we stand, how we came to be here, and what steps will perpetuate the spread of order that only recently engulfed us are nuggets of information that are now ripe for consideration, but only after eons of cosmic evolution laid the foundation from which we could arise. Now we, the products of natural undirected selection, must assume a more active directing role if the lifestream we belong to is to persist beyond the sun's expiration.

Our role as observers may have crucial significance by assigning some definiteness or reality to the universe. Would there have been a universe if nothing existed to notice it? If so, why? How would its existence differ from nonexistence? Could nature have a use for us? Was its evolution destined to include us, or at least some variety of intelligent observer? Prior to this century, these questions would have been judged absurd. Space, time, and matter/energy existed whether or not anything noticed them. And anything that did take notice would detect precisely the same values for all parameters. But two major developments of the twentieth century shattered this independence between observer and observed. The first, relativity, convinced us that lengths, times, and masses depend upon the state of motion of the observer. Lengths shrink, times slow down, and masses increase as one travels faster, at least according to a stationary observer. There is a difficulty deciding who is stationary and who is moving, but the main point is that the universe actually appears different to observers in *relative* motion. It is not the case that either errs in his measurements. At the extreme, when a mobile observer hitches a ride on a light beam (a feat he can accomplish only if he is massless), time stops altogether and length intervals shrink to zero. In other words, light that reaches us from a distant galaxy reveals

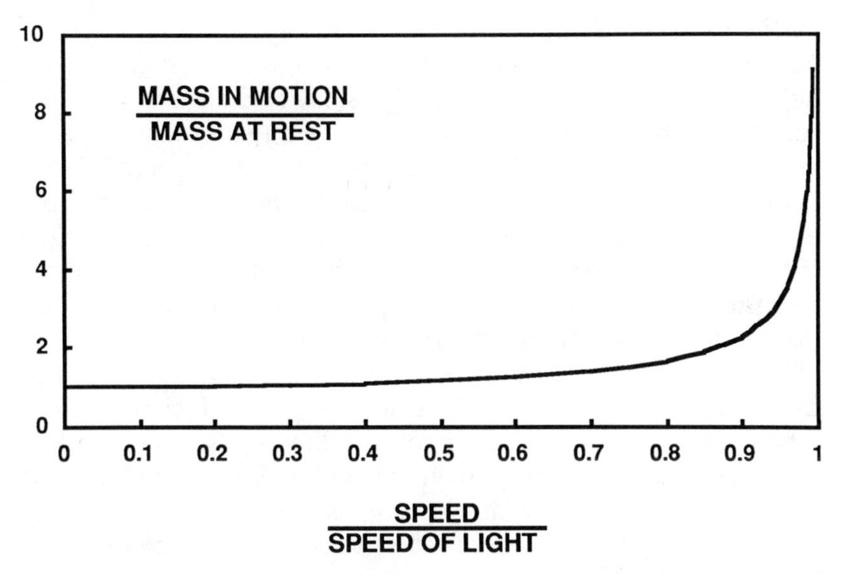

FIG. VII-10. Relativity. Masses depend on the relative motion of their measurers. A body moving at ninety-eight percent of the speed of light is five times more massive than when at rest.

its source as it was millions of years ago, but, as far as the transiting light photon itself is concerned, it left and arrived at the same instant.

The second scientific revolution of this century, that of quantum mechanics, introduced fundamental, inescapable uncertainties into our abilities as observers. A precise picture of the instantaneous state of the world is unattainable no matter how carefully we study it. It follows that the future can be predicted only probabilistically, not definitively. The histories and futures of individual fundamental particles—electrons, protons, photons, and the like—cannot be followed, since all are indistinguishable. The best we can do is to repeat the same observations of one type of particle sufficiently often that we gain a statistical summary of its probable behavior when subjected to a certain physical action. An inherent indefiniteness clouds our ability to specify the universe. And when we attempt to do so, we, by the very act of our measurement, force the physical system under study into a particular state; that is, our intervention has helped determine the subsequent history of the system. In that sense, the observer is no longer a passive outsider who watches reality unfold; he now influences some of the developments.

Does the universe transpire as a jumble of overlapping probability distributions until an observer forces a particular choice? If so, our value to nature is as great as hers to us. Or is there an infinity of universes, all simultaneously coexisting but completely independent, representing every possible outcome of every ambiguity, and ours just happens to be one path through the maze? Then we would seem creatures of circumstance whose role was not critical, because following the other fork at every branch point would have created an equally rich cosmic history. Or do our minds deceive us? Does reality lack any pattern or regularity except that which our limited mental abilities assign it? Note that even our minds operate in the dual modes of predictability and uncertainty, order and chaos. Electroencephelograms (EEGs), for example, reveal broadly similar patterns of electrical activity in the brains of many individuals, but no two EEGs are identical, not even ones taken from the same individual at different times. Individual neurons are activated randomly and spontaneously; networks of neurons form, dissolve, and reform; but the overall result is still (usually) orderly. Orderly, yes, but precisely predictable, no. Further, brain functions are at one and the same time localizable—for example, the left side usually controlling speech—and holistic—all of the brain functioning as a unit. The brain itself, as well as the universe it seeks to understand, both have built-in ambiguities. Is it any wonder our role is difficult to discern?

The questions raised in the last few pages are at present unanswered. The most conservative and least anthropocentric interpretation of the human predicament is, under the circumstances, the most defensible. According to it, we are merely convenient packages into which order (or information) has organized itself so that it can be spread more widely. We are assuredly not final steps: nature has none, apart from total extinction. We are merely transitions between organisms whose activities are rigidly proscribed by genes and those whose potential is limitless. The gift of creativity, our most precious asset, represents the shrewdest investment life can make to ensure its growth and continuity. Time, nature's ultimate filter of the unfit, will determine whether we were equal to the task.

For Further Reading

Diamond, Jared M. 1982. Evolution of bowerbirds' bowers: Animal origins of the aesthetic sense. *Nature* 297: 99–102.

Diamond, Jared M. 1982. Rediscovery of the yellow-fronted gardener bowerbird. *Science* 216: 431–434.

Diamond, Jared M. 1984 June. The bower builders. *Discover* 5(6): 52–58.

Hoagland, Mahlon B. 1978. *The Roots of Life*. Boston: Houghton Mifflin Company.

Luria, S. E. 1973. *Life, the Unfinished Experiment*. New York: Charles Scribner's Sons.

Pruett-Jones, Melinda, and Pruett-Jones, Stephen. 1983 September. The bowerbird's labor of love. *Natural History* 92(9): 48–55.

Restak, Richard. 2 November 1986. The endless levels of the mind. *The Washington Post*, p. C3.

Russell, Dale A. 1981. Speculations on the evolution of intelligence in multicellular organisms. In *Life in the Universe*, ed. John Billingham, pp. 259–275. Cambridge, Mass.: The MIT Press.

Rybczynski, Witold. 1983 September-October. The shock of the new. *The Sciences* 23(5): 22–26.

Stebbins, G. Ledyard. 1984 May-June. The flowering of sex. *The Sciences* 24(3): 28–35.

EIGHT

The Roots of Life

The continuous nuclear explosion that is a star came to earth and visited the eagle and the hemlock with energy made benevolent by the length of its journey. Their kinds were born of that energy and forever linked to it. The human co-planeteers of the eagle and the tree can never forget that their subjects are, like themselves, linked to a cosmos. That is the larger truth of the tree and the eagle, their link with everything that has ever been, everything that is, and everything that can ever be. . . . He who would strike the bird or the tree or man strikes at a child of the sun, and the anger of that parent is beyond the farthest edge of the human mind.

Roger Caras

LIFE IS A WEB of interacting cycles integrated into a whole whose performance cannot be predicted from knowledge of every individual component. The web of life continuously adjusts to maximize the efficiency with which it utilizes a flow of energy to impose the spread of order. Two fundamental questions are raised by the presence of life: (1) What is the source of the energy driving its quest for order? And (2) whence came the raw materials upon which life drew? The answers require that life's web be extended backward in time to an era before the solar system was present. They demand an understanding of stars.

As with life, a key point in stellar physics is the existence of cycles. Stars exhibit different behaviors at different stages during their existence. Their similarity to living systems is striking: one can identify stages of birth, maturation, and death. One can even assign a "stellar metabolism," a pace, to their activities. As with living organisms, stellar death is both an end and a beginning: the death of some stars can trigger the birth of others. Also, the birth of a few can occasion the birth of many.

Another analogy between living systems and stars is that neither can be divorced from their environments. In the case of a star the environment is always a galaxy. Gaseous and dusty material from the interstellar medium is drawn into stars, where it is chemically processed, usually for billions of years, and finally returned, at least in part, to its place of origin in interstellar space. When cycling into and out of stars, the matter of the galaxy behaves like that of the biosphere, whose materials cycle and recycle through organic and inorganic contexts. The timescales for clocking any atom's cycle are as disproportionate as the dynamics of the systems they belong to. In the case of atoms in earth's biosphere, the appropriate reference period is the planet's orbital course around the sun, the year; for an atom in the galaxy, the corresponding orbit is a star's around the galactic center, some hundreds of millions of times longer. Nevertheless, the principles of atomic recycling are analogous.

Since a loop has no well-defined starting point, we shall arbitrarily begin at the stage where stars spend most of their lifetimes. This is the stage in which the pressure of the radiation generated within the star balances its self-gravitation. Since the total gravitation of any collection rises with its mass, large stars must generate more radiation to resist it than small stars. In fact, the relationship between mass and luminosity (or radiated power) is a smooth one, commonly displayed on a Hertzsprung-Russell diagram (named after the astrophysicists who first plotted it). When stars are positioned on it according to their luminosities and surface temperatures, most fall along a narrow band called the main sequence. The parameter that determines where along the sequence a star belongs is its mass: the hottest, most luminous stars are the most massive, while the coolest, dimmest stars are the least massive. The sun is comfortably middling.

The luminosity or brightness of a star reveals how rapidly it is consuming fuel and can be considered a measure of stellar metabolism. The most luminous stars burn fuel much more rapidly than their greater mass compensates for, with the result that they have very short lifetimes. Those tens of times more massive than the sun live (i.e., generate energy) for periods measured in mere millions of years, compared with the hundreds of billions of years available to stars having only a few tenths the sun's mass, or with the ten billion years for a solar-mass star.

Maintenance of a stellar steady state on the main sequence is possible because of the resumption locally of a process performed everywhere for a short time in the early minutes of the universe, namely, the process of nucleosynthesis, the building of heavy elements from light ones. The initial rapid expansion

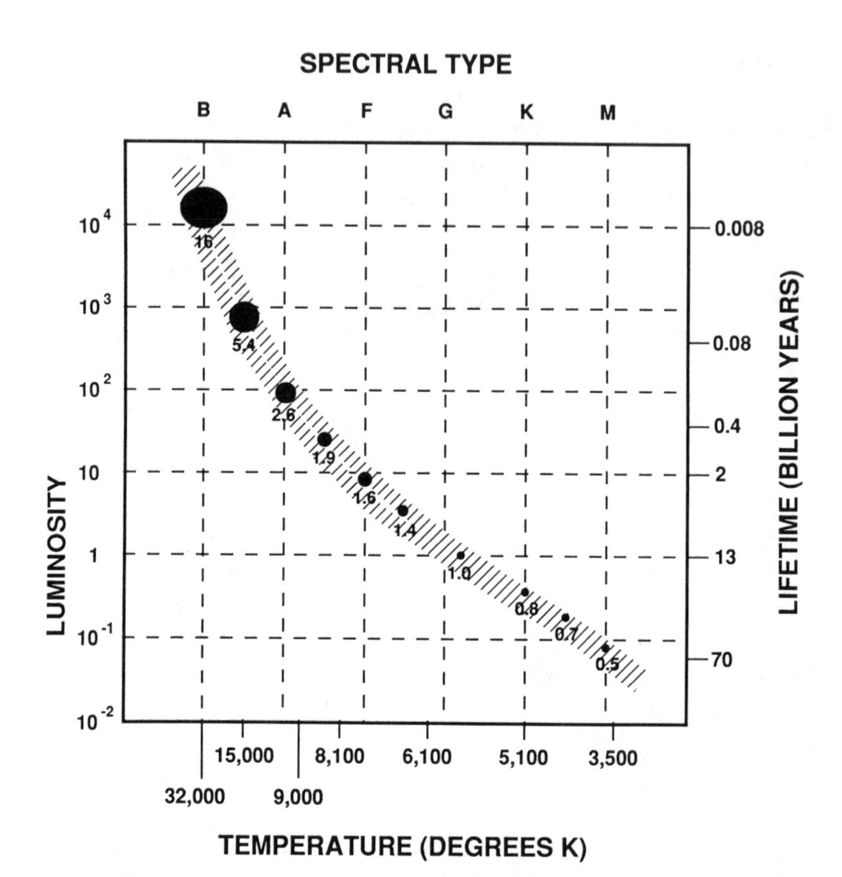

Fig. VIII-1. Hertzsprung—Russell Diagram. If stars' luminosities and surface temperatures are plotted, most occupy the band shown. The brightest stars are the hottest, but "live" the shortest lifetimes. Numbers near typical stars are their masses relative to the sun's.

of the universe and the slow pace of the weak nuclear interaction had conspired to halt element synthesis in the early universe far short of matter's highest entropic state, iron (cf. Figure IV–5). Most matter remained in the form with greatest potential for nuclear fusion, that is as atoms of hydrogen. Billions of years had to elapse before matter accumulated into aggregates sufficiently dense and hot to permit nuclear fusion. The appropriate aggregates did arise, though, as stars, places where some of the energy the universe held in reserve in the available (or potential) form of light nuclei is wrung out by fusing them into heavier ones. Throughout a star's main-sequence lifetime, the fusion at its center of four atoms of hydrogen into one of helium accounts for the outward flow of energy that counterbalances gravity. The energetics of this conversion have been presented in Table III–1. Only a small quantity of energy accompanies the production of a single helium nucleus, but in the case of the sun, 600 million (6×10^8) tons of hydrogen are being fused into 596 million tons of helium *every second*! Thus the energy equivalent of 4 million tons of matter radiates from the sun every second, an adequate flux to sustain life on a planet, Earth, 150 million kilometers from it.

From what kind of environment do stars emerge? Obviously one in which matter is concentrated to a high density, and one where great quantities of it reside. Otherwise the compression achieved by gravity can never accelerate protons (hydrogen nuclei) to speeds sufficient to overcome their mutual electrical repulsion. The matter must also exist in a cold environment; otherwise the internal pressure generated by random thermal motion disperses aggregations before they can trap themselves gravitationally. A region can be kept cold by preventing heat and light from entering it. Interstellar dust performs exactly this function. A concentration can also be colder if its membership is molecular rather than atomic. Molecules, electrically bonded atoms, are most likely to form where density is highest.

The necessary environments—cold, dusty, dense, and massive—are created when a density wave ripples through the disk of the galaxy. The wave itself could be triggered by the passage of a neighboring galaxy—or anything that perturbs the gravitational field holding a galaxy together. The density wave piles some matter up into arms emerging from the galaxy's center and leaves interarm matter dilute. The wave propagates like sound through air. Matter stays more or less in the same place,

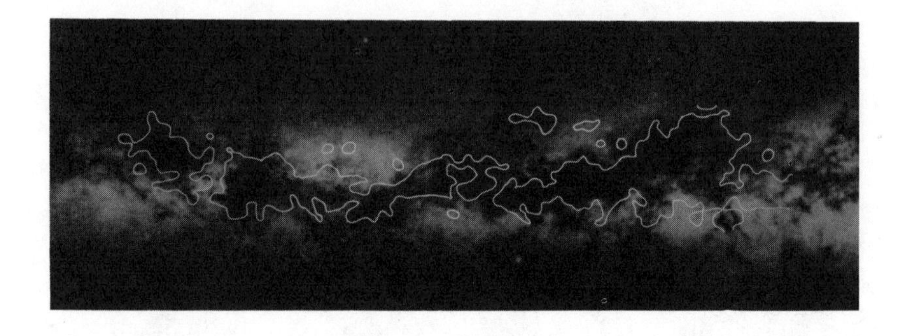

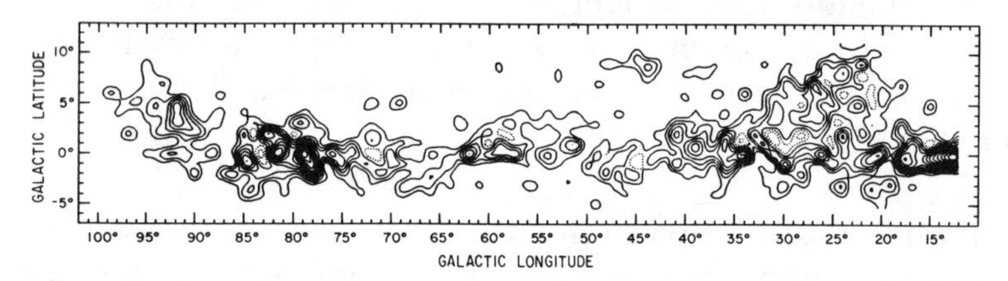

FIG. VIII-2. The distribution of carbon monoxide (CO) along the plane of the Milky Way Galaxy. The outer boundary of the contour map of CO density (bottom) is superimposed on an optical photograph of the Milky Way (top) to show that the molecule is concentrated where the sky is darkest (i.e., dustiest). (T. M. Dame and P. Thaddeus, Harvard-Smithsonian Center for Astrophysics.)

but is periodically compressed, then dispersed. During its compression phase, giant complexes of molecular clouds form along a spiral arm. These are proper settings for stellar nurseries. Thousands of these complexes have been detected despite their opaqueness to visible light. The molecules within them betray their presence by leaking telltale radio signals, at frequencies specific to each kind of molecule, to their surroundings. The clouds are hundreds of times denser than the average interstellar medium and contain hundreds of thousands of times the mass of the sun. They extend for hundreds of light-years. All told, they constitute the largest and most massive objects in the galaxy.

When, in a subsequent chapter, we consider the prospect of

extraterrestrial life, it will be useful to recall the variety of molecules radio astronomy has revealed. Many are organic. Their distribution is widespread and occurrence frequent. While they do not prove the existence of life elsewhere, they do indicate that chemical precursors to biologically active molecules are ubiquitous. Just compare the list of molecules detected in the

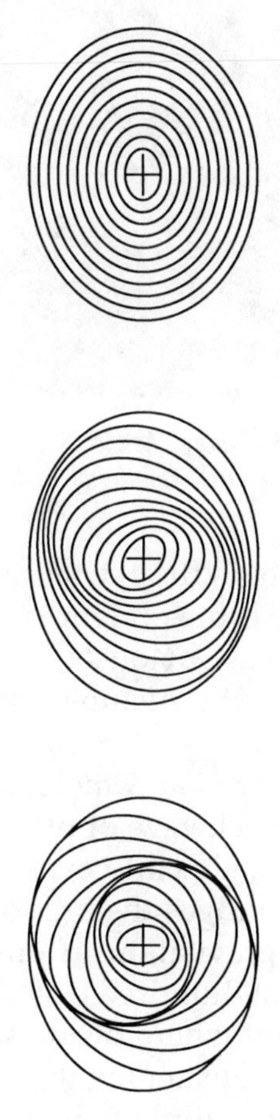

Fig. VIII-3. Density waves. Suppose stars move on elliptical orbits about the center of a galaxy. If something—the passage of another galaxy perhaps—rotates the axes of the ellipses at different rates, spiral patterns can build up.

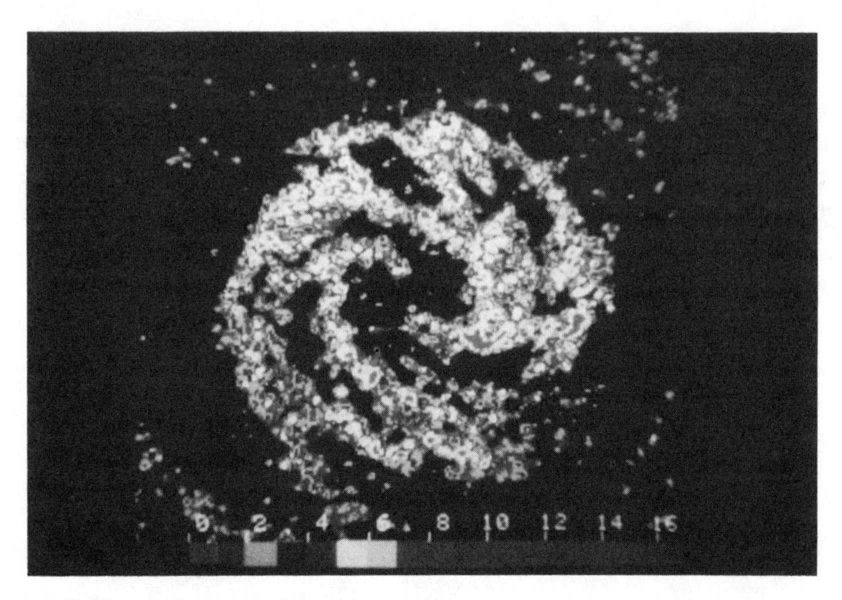

Fig. VIII-4. Radio Image of the Galaxy Messier 83, or NGC5236. Neutral hydrogen gas is compressed into spiral patterns. (National Radio Astronomy Observatory.)

interstellar medium (Table VIII–2) with the molecular structures of crucial biological molecules (Figures V–4 and V–5) to convince yourself of this fact. Obviously some of the steps toward assembling components of structures that constitute living organisms have been taken.

What happens if a star near a molecular cloud explodes, as many do at death? A shockwave of pressure advances through the complex, plowing matter into even denser configurations. This external stimulus nudges the preexisting balance between gas pressure and self-gravitation in gravity's favor. Collapse quickly follows, in just hundreds of thousands of years (a cosmic jiffy), to the point where fusion starts. At that moment, a star is born, capable of sustaining its structure against further contraction as long as its supply of hydrogen lasts. At first, the light of the young star will not be seen, so thick is the dust surrounding it. But both the heat radiating from the stellar sur-

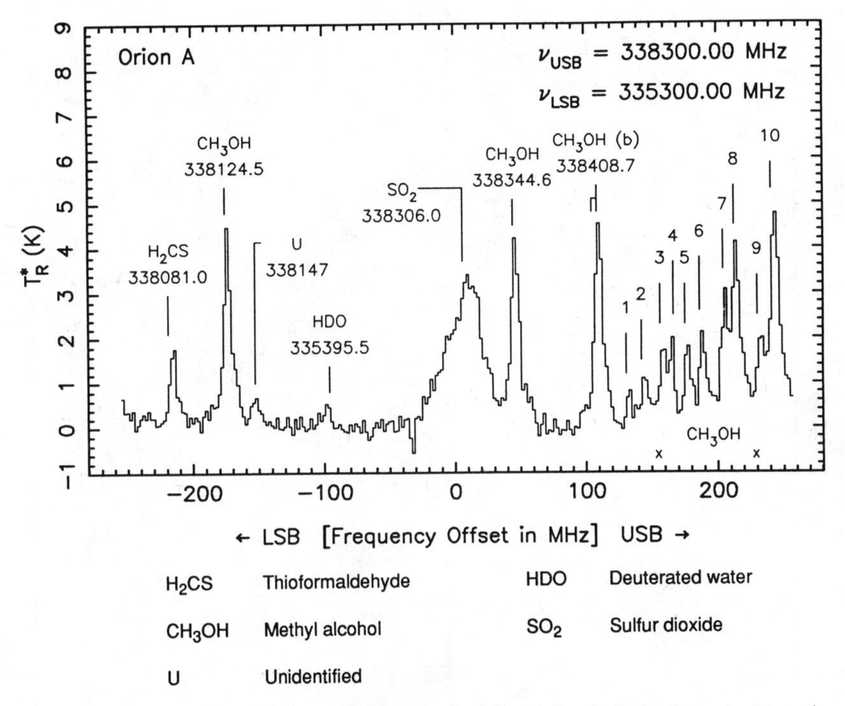

FIG. VIII-5. A portion of the radio spectrum of the Orion Nebula. Prominent peaks occur at frequencies characteristic of specific molecules. (Courtesy P. R. Jewell, J. M. Hollis, F. J. Lovas, and L. E. Snyder.)

face and the particles evaporating from it push outward the cocoon of its birth. Soon a glowing envelope of heated dust is seen by distant observers in the form of an infrared signal. The radiation emitted by the star and passing through its surround-

TABLE VIII-1. Parameters of Giant Molecular Clouds

Size	50–300 lightyears
Mass	10^5–10^6 solar masses
Density	100–300 cm^{-3}
Temperature	7–15° K

TABLE VIII-2. *Detected Interstellar Molecules*

# Atoms	2	3	4	5	6	7	8	9	10	11	13
Inorganic	H_2	H_2O	NH_3								
	OH	H_2S									
	SO	N_2H^+									
	SO^+	SO_2									
	SiO	HNO									
	SiS	NAOH(?)									
	NO	H_2D^+									
	NS										
	HCl										
	PN										
Organic	CH^+	HCN	H_2CO	HC_3N	CH_3OH	HC_5N	$HCOOCH_3$	HC_7N	CH_3C_5N(?)	HC_9N	$HC_{11}N$
	CH	HNC	HNCO	C_4H	CH_3CN	CH_3CCH	CH_3C_3N	$(CH_3)_2O$	$(CH_3)_2CO$		
	CN	C_2H	H_2CS	H_2CNH	CH_3SH	CH_3NH_2		CH_3CH_2OH			
	CO	C_2S	HNCS	H_2C_2O	NH_2CHO	CH_3CHO		CH_3CH_2CN			
	CS	SiCC	C_3N	NH_2CN	H_2CCH_2	H_2CCHCN		CH_3C_4H			
	CC	HCO	C_3H(lin)	HCOOH	C_5H	C_6H					
		HCO^+	C_3H(ring)	(CH_4)							
		HOC^+(?)	C_3O	(SiH_4)							
		OCS	C_3S	C_3H_2							
		HCS^+	$HOCO^+$								
		CCS	(HCCH)								
			$HCNH^+$								

Atomic Symbols:
H = Hydrogen O = Oxygen P = Phosphorous C = Carbon Na = Sodium S = Sulfur N = Nitrogen Si = Silicon Cl = Chlorine

ing dust cloud can trigger the formation of other stars. It does so by piling matter up at its advancing front. Often, then, hot, young stars are seen in groups.

As the wave of star formation advances through the cloud, the latter dissipates, eaten away by the new stars forming within it and blown apart by their intense radiation. The material escaping that round of star formation resides in the disk of the galaxy until another density wave renews the cycle of activity.

A short summary may help the reader absorb the key facts about the birth of stars. The process starts when a density wave propagates through a galaxy, producing spiral arms denser than the interarm regions. Giant Molecular Clouds (GMCs) congregate within the spiral arms. Occasionally a star within or near a GMC explodes. The shock wave advancing from the explosion site compresses matter into collapsing protostars. The collapse of a protostar is halted when its internal temperature and density rise to the point where nuclear fusion can begin. The energy released by fusion not only keeps overlying layers of the star from falling inward; it also pushes back out into the GMC, where its advancing pressure front may trigger additional star births. The GMC dissipates into the galactic disk until another density wave initiates another cycle.

One accompaniment to star formation deserves mention. Gas driven away from hot, luminous young stars usually emerges in oppositely directed flows. Indirectly, these are evidence for disks of matter centered on the stars, but hollowed out near their centers (visualize a doughnut with a star in the center of the hole); for if a such disk is present, matter blown off from the star escapes easily through the disk's poles, but meets stiff resistance around its equator. If such disks are common about stars, then planetary systems are likely to be numerous. Indeed one in its formative stages may already have been spotted around the star β Pictoris (see Figure X–7).

What precipitated the death of the star in the giant molecular cloud, an event that triggered the birth of several others? The answer is exhaustion of the fuel for fusion and the consequent cessation of the flow of energy from the spent star's center. When this outflow ceases, the infall gravity has been trying to accomplish since the star's birth transpires. The complete exploitation of gravitational energy—held in reserve since the initial rapid expansion phase of the universe—is finally accomplished. The potential energy matter possesses when it stays in a configuration that allows it to fall is converted totally into kinetic energy and heat. The star's matter falls to the bottom of a gravitational well where it lies degraded to a form from which no further useful work can be extracted— like water at the base of a waterfall. If the dead star's mass had been several times that of the sun,[1] its collapse would have been total, the entire core shrinking to a density such that its gravity becomes too powerful for even the fastest-moving particles, photons of light, to escape. It is as though all of space and time near the remnant has sunk into a well of infinite depth. These space-time singularities are called black holes. As the huge star turned into a black hole, inwardly collapsing matter had to overpower two major forces. The first arose when electrons of similar energy were crowded into near coincidence. The rules of atomic physics forbid such a circumstance. These rules manifest themselves in the electronic structure of the various kinds of atomic elements: only certain discrete energy levels (or orbits) about the nucleus are allowed. But this so-called "electron degeneracy pressure" can be overcome if the compression is strong enough. If it is, electrons and protons are squeezed together to form neutrons. Neutrons, though, also resist sharing the same space if they have the same energy, and their resistance is even fiercer than electrons'. This "neutron degeneracy pressure" also had to be overcome for the matter to fall into a black hole.

The two pressures opposing gravity hint that stars with in-

[1] The mass of the sun is about 2×10^{33} grams. This is 333,000 times the mass of the earth. It is approximately the middle of the range of stellar masses.

sufficient matter to fall all the way into black holes could have different final resting states, and indeed they do. The lightest stars—those at most a few times more massive than the sun—cannot overcome the electron degeneracy pressure. They solidify into structures resembling gigantic atoms, dense but not infinitely so, called white dwarfs. Stars of intermediate mass—between a few times and an order of magnitude more massive than the sun—terminate with the density of matter in an atom's nucleus. They are called neutron stars.

In all three terminal cases, white dwarfs, neutron stars, and black holes in order of ascending mass and density, not all of the stellar matter is locked in the remnant. Some fraction of the gravitational energy released by the core's rapid collapse is transferred to the overlying layers. There, the energy manifests itself partly as visible light, partly as motion of the debris, and partly as cosmic rays, extremely energetic subatomic particles. Depending upon how rapidly the core collapses, the resulting eruption of the outer stellar matter can be as gentle as a smoke ring (a planetary nebula) or as violent as a thermonuclear explosion (a supernova). Central to the existence of life is the fact that the matter being returned by stars at death to the medium that spawned them is chemically enriched. Otherwise, living things would have to be made only of hydrogen and helium. But the chemical bonds between these two atomic "lightweights" are too weak to link chains of molecules sufficiently long to contain the information needed for "earning a living."

TABLE VIII-3. *Stellar Corpses*

Mass of Original Star	Mass of Collapsed Core	Type (Solar Masses)	Diameter (Kilometers)
<1.5	<1.4	White Dwarf	10^4 (Planet size)
3–8	2–3	Neutron Star	20 (City size)
>8	>3	Black Hole	>10 (Horizon only)

Fig. VIII-6. The Helix Nebula, a planetary nebula. The dot at the center of the ring is a white dwarf, the core of a star that died gently. That star's outer layers are expanding into interstellar space at a speed of tens of kilometers per second. (National Optical Astronomy Observatories.)

Moreover, a two-element chemistry set offers little possibility of variety. Fortunately, stars systematically fuse enough heavier elements to permit diversity, some of which, most notably carbon, can also provide bonding strength.

A star's collapse toward finality proceeds in stepwise fashion. Once the interior hydrogen is exhausted and replaced by its helium residue, the inward pull of gravity raises central temperatures and densities to levels where helium fusion can begin. The product of this nucleosynthesis is carbon. It accumulates at the expense of helium until the supply of the latter is entirely consumed. Further collapse and heating then resume until carbon can be synthesized into oxygen. On and on, up the table of periodic elements, at least as far as iron. Beyond that, chemical diversity depends upon the violence of supernovae explosions. To future generations of life these spew out calcium for bones, iron to be carried by blood, and carbon for cells. That the solar system contains life is proof that stars lived and died before its formation.

Meteorites deliver a message that more happened before the birth of the solar system than the mere production of chemical elements. Meteorites are bits of matter that were present in the nebula out of which the bodies of the solar system formed, but that did not agglomerate into any of them. They therefore preserve cosmic history from epochs predating the solar system. Occasionally a meteorite large enough to escape burning up in earth's atmosphere strikes its surface. When chemically ana-

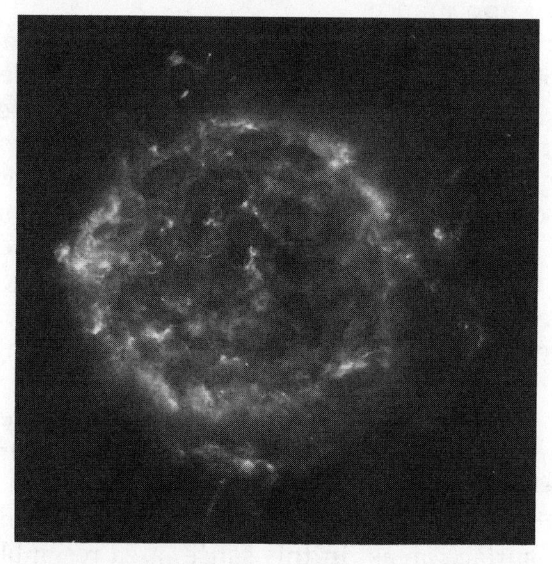

Fig. VIII-7. Radio image of Cassiopeia A, a supernova remnant. This star's death was violent, matter being expelled outward at several thousands of kilometers per second. Moreover, the dispersing matter has all the chemical elements, some fused during the star's lifetime, but others created in the explosion itself. (National Radio Astronomy Observatory.)

lyzed, it is found to contain amino acids, not just heavy hydrocarbons (present as well), but actual amino acids, arguably the most important building blocks of living organisms. If amino acids were present in the nebula the solar system condensed from, then they are probably present in other similar systems as well. The cosmos may therefore be a surprisingly fertile breeding ground from which life can spring—not just once but often.

Since gravity is both the consequence of the matter and radiation that occupy the universe and the shaper of their distributions, its ultimate repository in black holes is central to the timeline threading the entire story we are revealing. Black holes loom prominently in our future, however remote, since it is obvious that thermonuclear fusion cannot indefinitely postpone their appearance.

Matter in any quantity can be locked in a black hole. It need merely be packed so compactly that the escape velocity at its surface is at least equal to the speed of light, 300,000 kilometers per second (186,000 miles per second). Its surface at that exact boundary is called the event horizon. Of course, the size of the material concentration at the requisite density depends directly upon its mass. The earth, for instance, would have to be compressed to a radius of about one-third of an inch (from its present 251 million inches) to become a black hole.

Surface area is a better parameter for specifying the event horizon, since space near a black hole is severely warped. As a result, spheres are no longer basketball shaped. Whenever matter or radiation is added to a black hole, the surface area of its event horizon increases. Moreover, when two black holes merge, the resultant surface area exceeds the sum of the two entering surface areas. In this sense, area resembles entropy. Recall that entropy is the property physicists use to describe the degree of order in a system. Highly ordered systems have low entropy, whereas chaotic ones have high entropy. Since the natural tendency of closed (i.e., isolated) systems is to decay

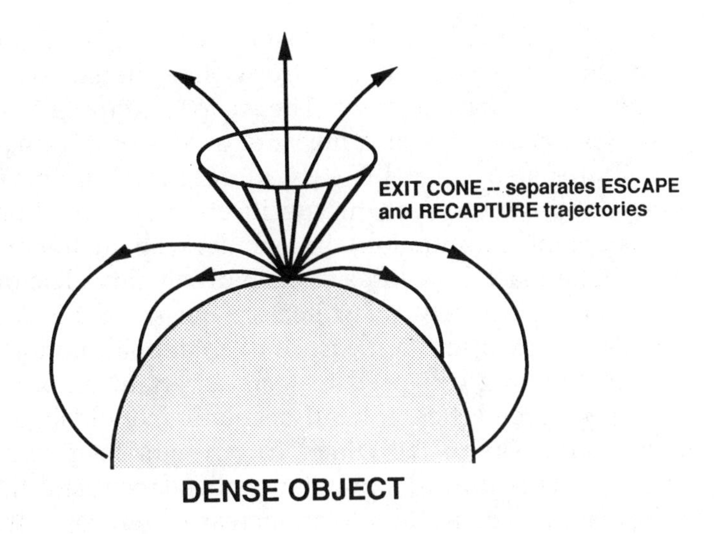

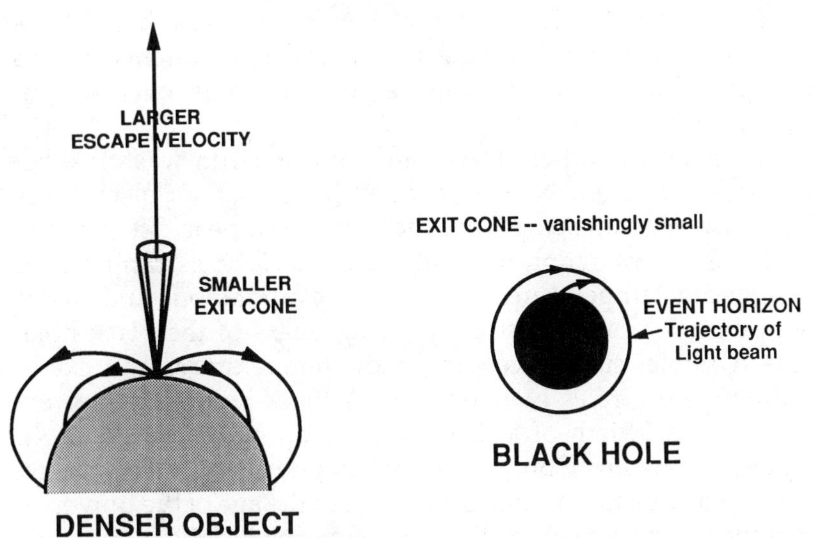

FIG. VIII-8. *As the density of an object increases, the cone within which a projectile, to escape, must be aimed shrinks, and the launch speed of the projectile increases. At a sufficiently high density, the launch speed reaches the unattainable speed of light. Nothing can escape then.*

from order into chaos (iron rusts, rocks fall, an odor diffuses throughout a space, etc.), entropy, like the surface area of a black hole, always increases. The analogy is strengthened when we consider entropy as a measure of loss of information. In a physical system, chaos is easy to describe (minimal information) but order requires the specification of several parameters (considerable information). In a similar way, matter or radiation entering a black hole loses *all* its individuality. The only properties not completely eradicated are mass, electric charge, and angular momentum (a measure of rotational motion or spin). The matter that collapsed to form a certain black hole could have been green cheese, hand calculators, or old jogging shoes; there is no way to tell which. So an enormous loss of information occurs during a gravitational collapse, and that loss of information is equivalent to an increase of entropy, the natural course of events. The fact that combined black holes have greater surface area than their individual areas summed means that the universe favors one large black hole over several smaller ones.

Matter on the other side of an event horizon was once believed lost forever. Now we know that is not so: black holes leak particles into space until the holes disappear. The process resembles evaporation of matter, and black holes can even be assigned a temperature. The rate of evaporation, and hence temperature, depend strongly on the mass of the black hole. Massive holes have exceedingly low temperatures and excruciatingly slow rates of mass loss. A black-hole earth, for example, at a fiftieth of a degree above absolute zero (0.02°K), would radiate away only after ten trillion trillion trillion trillion (10^{49}) years, some 10^{39} times longer than the age of the universe.

How can matter escape the firm grip of a black hole's gravity? Quantum mechanics, the system that describes the behavior of subatomic particles, provides the answer. It tells us that space is never truly empty. Particles and antiparticles (identical but for opposite electric charges) can spontaneously arise from nothing provided they quickly disappear. Nature, appearances

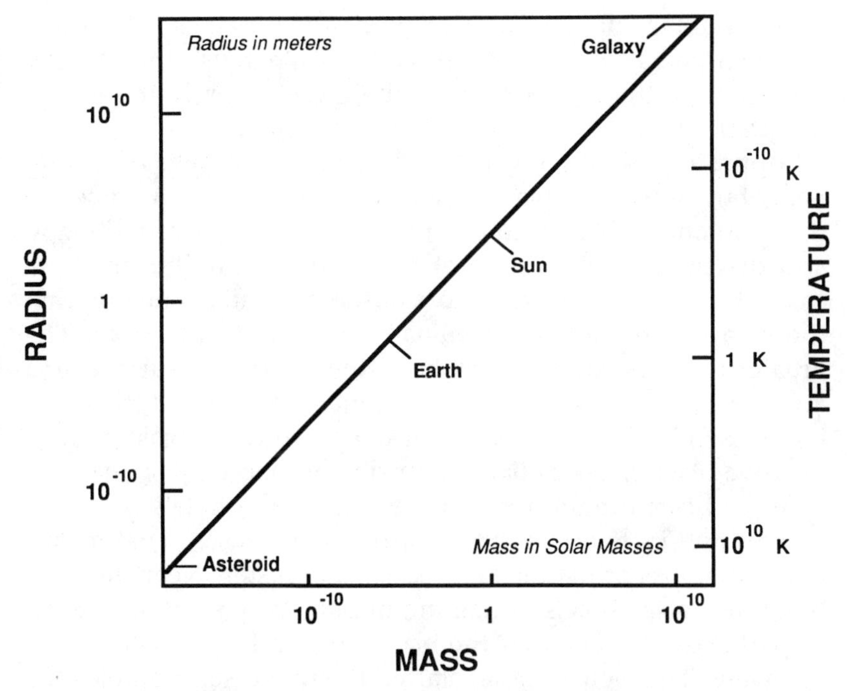

FIG. VIII-9. The radii of black holes grow directly with their masses; their temperatures decrease inversely with mass.

to the contrary, is not providing a "free lunch": the whole creation-annihilation cycle takes place behind the screen of uncertainty that prohibits arbitrarily precise measurements. Thus the energy *and* the lifetime of a physical system cannot both be measured exactly. The more precisely one is determined, the coarser becomes knowledge of the other. For example, if one knew precisely when a particle-antiparticle pair sprang into existence, he would be completely ignorant of their energies; correspondingly, precise knowledge of their energies precludes knowing when they existed. So a lot of action can occur—and is hypothesized to be doing so—beneath the "fog" created by our inability to measure with absolute precision. The situation is akin to having blurred vision, so that we never see some things that take place. A particle–antiparticle pair can "borrow" as much energy as it wants provided it "repays" that energy fast enough; the less borrowed, the more leisurely the repayment schedule.

Suppose, now, the invisible "frothing" of space is occurring near, but outside, a black hole. Occasionally one member of the spontaneously created pair might fall over the event horizon and disappear. The other member could follow, but in some cases it would escape. A remote observer would conclude that the escaping particle had emanated from the black hole. The greater the mass of the black hole, the greater its surface area and the stronger its surface gravity, making it more difficult for either member of a "virtual" pair to escape. This explains why massive black holes radiate more slowly than lighter ones.

As particles tunnel out of a black hole, the hole's mass decreases and its temperature accordingly increases. That makes additional loss of matter even easier and faster. Near the end, when the black hole's remaining mass is very small, emission of particles is so fast that the black hole ends its existence explosively. The big bang that started the universe resembles the explosion that ends the existence of an enormous black hole. Could we live in a universe whose matter and energy erupted from the total gravitational collapse of an equivalent quantity participating in an earlier phase of the universe? If so, all information from that previous phase is lost. Could big bangs have followed big crunches endlessly into our past? The possibility at least expands our notions of time. It also makes the present universe's existence less special by suggesting it may be but one cycle among an endless string. Could life have been a certainty sometime, simply because its ingredients are periodically reshuffled until appropriate conditions fall into place?

Finally, let us consider again how humans serve nature by observing her. We discovered during this century that quantum mechanics cut our predictive abilities in half. Exact knowledge of certain parameters came at the expense of total ignorance of

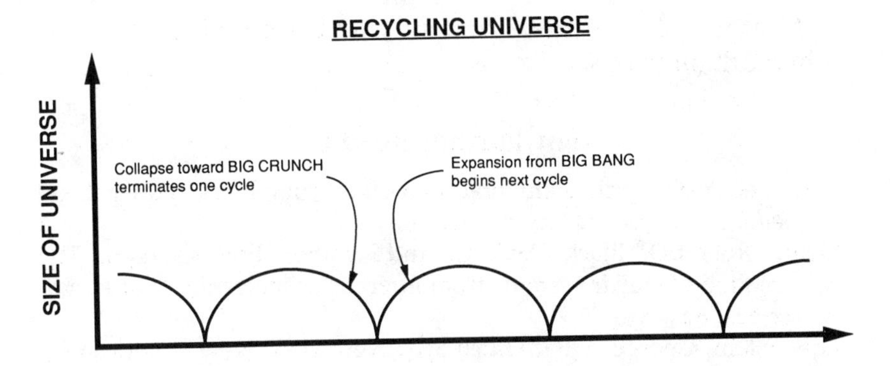

Fig. VIII-10. A recycling universe. An expansion cycle emerges from the big bang created by the big crunch terminating a collapsing cycle.

their complements. In this chapter, energy and time were complementary parameters; others are velocity and position, or spin and angle of rotation. If we knew either one precisely, we had total ignorance of the other, not because we were careless with our measurements but because the act of making them changed the quantity we were trying to measure. Black-hole physics diminishes our abilities still further. As entropy continues its inexorable increase, black holes of larger mass assemble. The particles they emit confound us not only by eluding precise specification, but also by originating in regions that contain very little information. Black holes erase the identity of everything that has fallen into them. Observers will never be certain what type of particle a black hole will emit, or when or where it will do so. At best they can predict probabilities. The crystal ball of the universe's future reveals only a fog of possibilities, not certainties. Nature, contrary to Einstein's oft-quoted assertion that "God does not play dice with the universe," seems to be doing just that. Worse still, it is doing so where the croupiers are hidden. The universe may indeed have needed to include a capacity for self-observation to acquire reality, but it will have to be content with a view of itself that is inherently imprecise.

For Further Reading

Blitz, Leo. 1982 April. Giant molecular-cloud complexes in the galaxy. *Scientific American* 246(4): 84–94.

Gehrz, Robert D., Black, David C., and Solomon, Philip M. 1984. The formation of stellar systems from interstellar molecular clouds. *Science* 224: 823–830.

Greenstein, George. 1983. *Frozen Star*. New York: New American Library, pp. 115–191, 244–266.

Hawking, S. W. 1977 January. The quantum mechanics of black holes. *Scientific American* 236(1): 34–40.

Hawking, Stephen W. 1988. *A Brief History of Time*. New York: Bantam Books.

Lada, Charles J. 1982 July. Energetic outflows from young stars. *Scientific American* 247(1): 82–93.

McClintock, Jeffrey. 1988 January. Do black holes exist? *Sky and Telescope* 75(1): 28–33.

Sullivan, Walter. 1979. *Black Holes*. New York: Warner Books.

Talcott, Richard. 1988 February. Insight into star death. *Astronomy* 16(2): 6–23.

Verschuur, Gerrit L. 1987 May-June. Molecules between the stars. *Mercury* XVI(3): 66–76.

Wald, Robert M. 1977. *Space, Time and Gravity*. Chicago: University of Chicago Press.

Waldrop, M. Mitchell. 1983 May. Stellar nurseries. *Science 83* (4): 40–47.

NINE

Immortality?

If my view of the future is correct, it means that the world of physics and astronomy is also inexhaustible; no matter how far we go into the future, there will always be new things happening, new information coming in, new worlds to explore, a constantly expanding domain of life, consciousness, and memory.

Freeman J. Dyson

AN ORGANISM is a complete cycle. Any discussion of life that does not include the time dimension is as restricted as a discussion of geological features in which elevation is omitted. Even the total life cycle of an individual organism is an incomplete sample of life, however, for an organism is too small a unit. No single organism can exist without the support of many others of various kinds, just as no single cell of your body can function in complete isolation. Rather, all life is a *collectivity.* On earth, it constitutes one superorganism. That superorganism has a time component, whose past we have already examined. Earth's life may presently be in a critical phase: the same abilities that enable humans to understand their superorganism and its place in the universe also enable them to destroy that superorganism. No evidence exists, however, that its growth, spread, and diversification *must* end soon, or even ever. In short, the cycle of the superorganism is incomplete; optimism and faith in its continued vitality demand that we ponder its future.

The future of life is ultimately decided by that of the universe. As long as some disequilibria persist in space and time, an orderly flow along natural gradients can be maintained: a few examples include the flow of heat from hot reservoirs to cold ones, the flow of matter from loose configurations to tight ones, and the "flow" of nuclei from light to heavy. At least in principle, any such ordered or directed flux can be transferred somehow into the reproducing, metabolizing, and mutating order we recognize as life. The potential for life is as enduring as the universe's ability to postpone degradation to a state of minimal order and information, at which condition all activity ceases.

How will the universe end? It has but two options: either it will continue its present expansion forever, or it will eventually reverse directions and undergo indefinite contraction. The former case describes an open, infinite (in both time and space) universe; the latter a closed, finite universe. For life, the corresponding options appear to be, respectively, death via a deep freeze or via cremation. Intelligent life may be able to avoid freezing, but not in all probability frying. Still, even the closed universe offers a slim chance that its collapse is not terminal in perpetuity, in the sense that space and time might rebound into a new cycle of activity (see Figure VIII–10, for example).

When we discussed the fate of the universe previously (see particularly Chapters II and IV), we determined that it lay close to the borderline situation between open and closed models. We cited this as evidence for a runaway expansion early in the history of the part of the universe we inhabit, a part we could never have inhabited if its space had not been close to flat.[1] How do we know the balance between the inertia of matter and the vigor with which it was launched outward is close to equal? Is there a tilt in favor of either gravity or expansion? We shall cite four clues slightly favoring expansion to infinity, not because the clues are definitive, but because they illustrate the methods for determining the universe's future.

First, we can look at distant galaxies. The light from them has been in transit for a long time. Consequently, it portrays the universe as it was at much earlier epochs. Were galaxies separating faster then than now? If so, we can gauge the deceleration with which all the matter of the universe brakes itself. Knowing that, we can extrapolate into the future to determine whether the braking is sufficient to reverse the present epoch of expansion. The evidence now says it is not, but the uncertainty of that evidence leaves alive the possibility of a closed universe.

[1] Flat space describes a universe that will expand forever, but at an ever-decreasing rate, until it reaches infinite dimension only after the passage of infinite time.

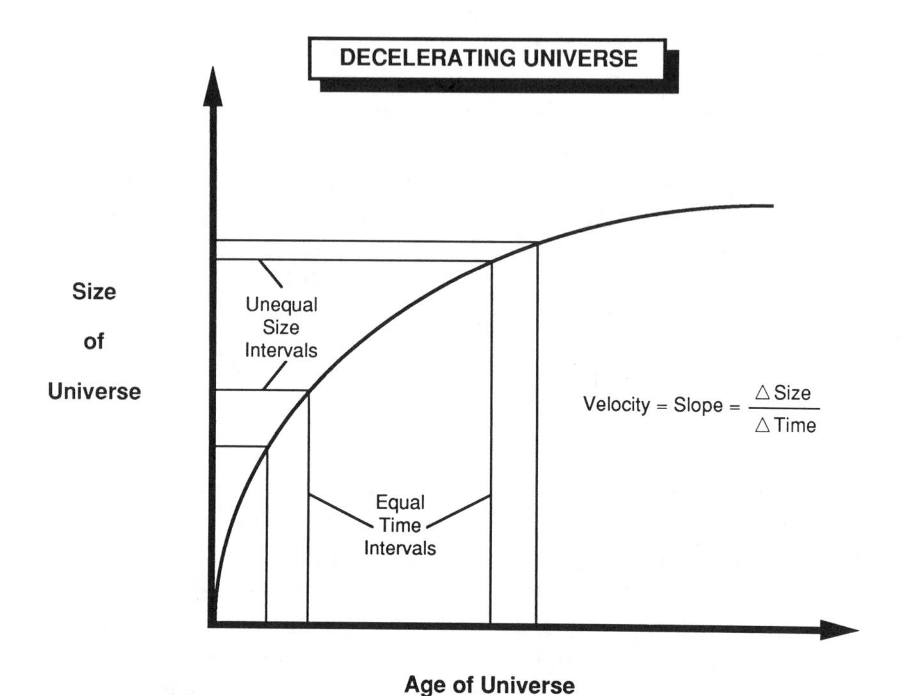

FIG. IX-1. In curves representing size of the universe vs. age, the slope at any moment represents the velocity of expansion. By measuring velocities at different epochs, one can tell how much deceleration has occurred. Distant galaxies reveal velocities at past epochs.

A second clue to the distant future emerges from the distant past. How much deuterium, cooked in the first few minutes following the big bang, has survived to the present? The answer depends very sensitively upon the mass density of matter during that era. As pointed out in Chapter III, deuterium is a difficult nucleus to form. Once formed, however, it is easily and rapidly combined into helium. It follows that, if the density of matter early on had been high enough to close the universe, nearly all of the deuterium created would have been immediately reprocessed into helium; if the density had been low enough to permit an open universe, some deuterium would have escaped the collisions with other particles that lead to further fusions. Some deuterium—one or two atoms in every hundred thousand—did indeed escape incorporation into heavier elements. The implication is that the primeval matter density, converted into a present density by calculating how much the universe has expanded since the era of nucleosynthesis, falls a factor of at least ten and perhaps as much as one hundred below the value necessary to halt the expansion of the universe. The trouble is that the calculation of conditions separated by so much time introduces enough uncertainty to weaken the conclusion.

When more accurate measurements and theories tighten the limits on the mass density of the universe, will space turn out to be precisely flat? Philosophically, the prospect is troubling. The post-Copernican scientific framework shuns assumptions of special properties even for a sample, in this case the universe,

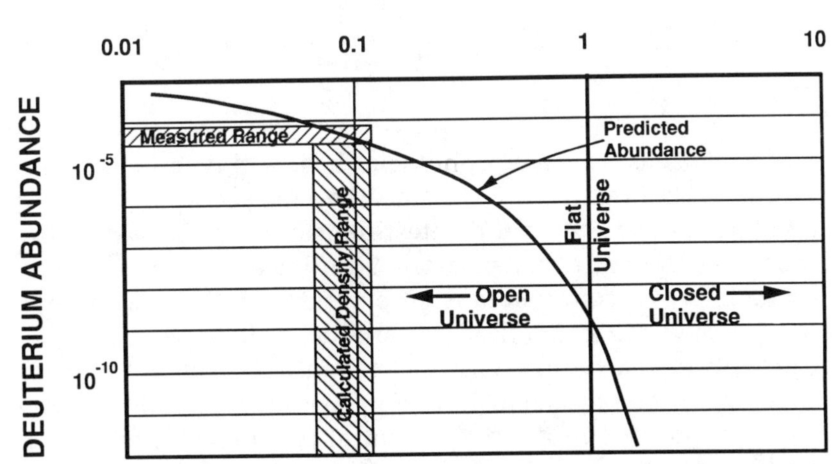

FIG. IX-2. Abundance of deuterium (relative to hydrogen) for various densities of the universe. The measured abundance of D/H indicates that the universe lacks the density needed for a flat space by about a factor of ten.

which, having a single member, is by definition unique. But the mass density is *extremely* close to the critical value: a change of a mere part in 10^{57} (a 1 followed by 57 zeroes, the approximate number of nuclei in the sun) means the universe either would have recollapsed long ago or would have expanded so fast that galaxies, stars, and planets could never have congregated. Is it likely that a balance so fine could have been struck without being precise? We are tempted to conclude not, but then we are driven to postulate a "missing mass." This is the amount of matter that must be added to that we can observe directly if the total is to be sufficient to decelerate the universe's expansion to zero after an infinite time. The designation is inaccurate, since evidence for the presence of "hidden matter" is abundant; that is, the matter is not missing, it is there. Whether in sufficient quantity to close the universe is as yet unknown. Speculation assigns it many forms—tiny but profuse black holes, extremely dim stars of low mass, large planets, subatomic particles once thought massless—but until some glimpse of its nature is gathered, the hypotheses are unbounded. Most embarrassing of all, the invisible mass is not a minor constituent of the universe. It may total ninety percent of all matter!

How can we be sure of its presence? Sometimes we see clusters of galaxies whose morphologies belie a gravitationally "relaxed" condition. That is to say, the cluster approximates a spherical shape with the greatest density of galaxies, and the most massive individual galaxies, found near its center. This is an equilibrium configuration for a system of bodies that have had ample time to interact gravitationally. It comes about because every galaxy that started in a highly elongated orbit feels tugs from each of the many other galaxies it encounters during its passage through the dense central part of the cluster. The multiple tugs amount to a form of friction. They rob the galaxy of the energy it needs to return to the most distant part of its orbit, far from the main matter concentration. It follows that after several such passages, the orbits begin to approach circularity. The point is that the appearance of a cluster reveals whether or not it is a gravitationally bound system. The Coma Cluster (Figure II–2) has the appearance of a gravitationally bound system. Surprisingly, however, the galaxies in Coma are moving too rapidly to be held together by the amount of matter that can be seen. There must be a nonluminous component, ten or twenty times more massive than the luminous one, binding the cluster together. Even this amount of hidden matter is inadequate to close the universe, however. Yet a third clue points to an unbounded space-time.

The location of at least some of the invisible matter is known. Measurements of the rotation of spiral galaxies indicate that dark (i.e., nonradiating) matter extends well beyond the visible outlines. In centrally concentrated systems of mass, the velocities necessary to retain a body in orbit decrease as one moves out from the center. The velocity determines the body's centrifugal force, and this can decline in concert with the gravi-

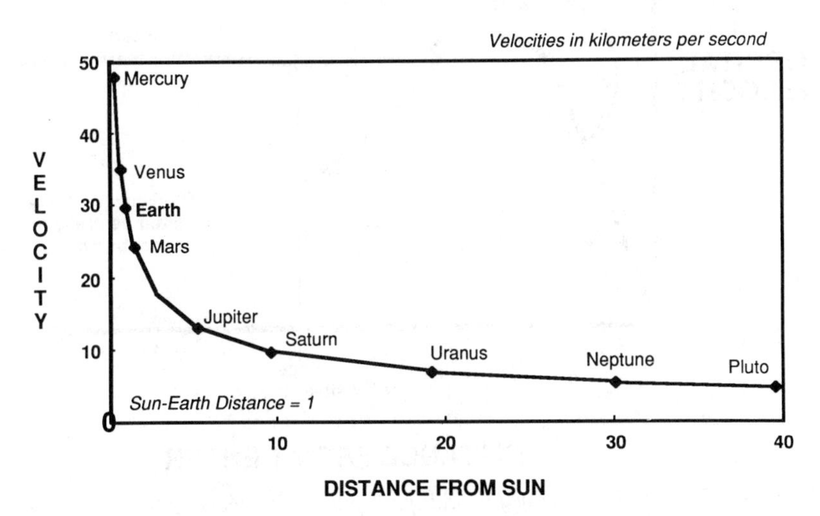

FIG. IX-3. Orbital velocities of planets. More than 99 percent of the solar system's mass is concentrated in the sun at its center. The velocities needed to stay in orbit about this mass decrease with distance from it.

tational force's falloff with distance from its main source. The principle works for satellites in orbit about the earth and for planets orbiting the sun. It would also work for stars in a galaxy—*if* the galaxy's mass were confined near its center. In every well-studied case, it is not.

Orbital velocities in galaxies first rise rapidly as the orbits get farther from the center. This reveals that the matter in the galaxies' inner portions is rotating like a solid body: the end of a stick spinning around its short axis rotates twice as fast as its midpoint, for example. But eventually—slightly inside the sun's orbit in the Milky Way—the orbital velocity reaches a maximum. Instead of falling at greater distances from the galaxy's center, as the distribution of luminous matter predicts, the orbital velocity stays near its peak value way out to the edge of the visible matter—and beyond, if we use radio signals to

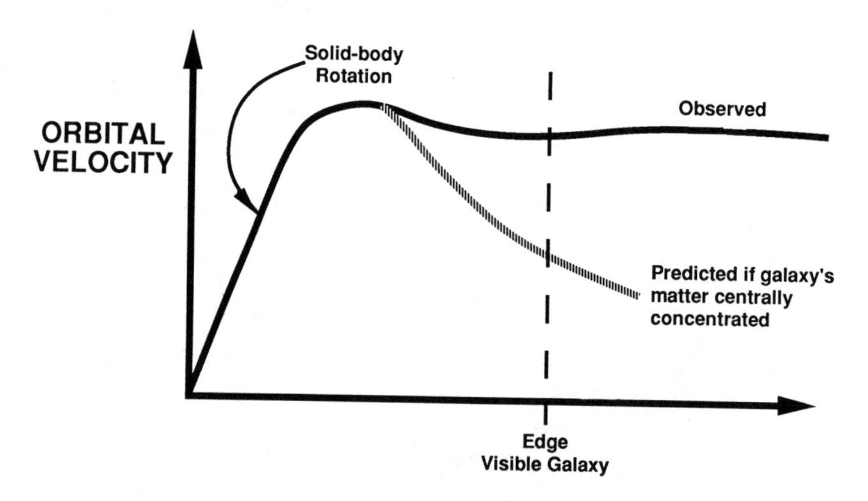

FIG. IX-4. The rotation curve of a typical spiral galaxy. Orbital velocities do not drop even at distances well beyond the visible edge of the galaxy. This indicates matter is not concentrated within that visible boundary, or the velocities would fall as they do in Figure IX-3.

probe optically dim matter. This is evidence that the mass is not tapering off; it is merely shifting its countenance from bright to dark. But even if we add the mass in their dark, enormous halos to that in the visible part of spiral galaxies, we still fail for a fourth time to find sufficient matter to close the universe.

The fact that much more matter cannot be seen than can shatters the confidence astronomers have in their ability to decide between open and closed, infinite and finite, eternal and limited, universes. As we consider ever larger structures in the size hierarchy of the universe, we approach ever more closely the amount of matter necessary to halt its expansion. In fact, the ratio of the mass of a configuration to its luminosity appears to increase approximately linearly with the scale size of the configuration, and the largest aggregations come tantalizingly close to the critical value. This sustains the believe that we live in a universe near the knife-edge of a balance . . . but perhaps it could not be otherwise.

An open universe offers to life the prospect of an unending future. A closed, finite universe, in direct contrast, has the depressing consequence of total, final, unavoidable termination for all life. The end, though, if indeed one comes, is not imminent; expansion of the universe should continue for another fifty billion years if not forever. Thereafter, if we hypothesize a closed universe, history will play itself out in reverse. In eighty or a hundred billion years, most stars will have been long dead, their supplies of fuel suitable for thermonuclear fusion completely spent. Galaxies will consequently be decidedly less luminous and will be approaching, rather then receding from, each other. When the hypothetical closed universe is one percent its present size (occupying a millionth of its present volume), galaxies will be in contact. The background radiation, remnant of the big bang, will be getting compressed and thereby heated. When the universe has contracted to one two-thousandth of its present size, the temperature of the radiation will be 5000° to 6000° Kelvin (approximately 10,000° Fahrenheit),

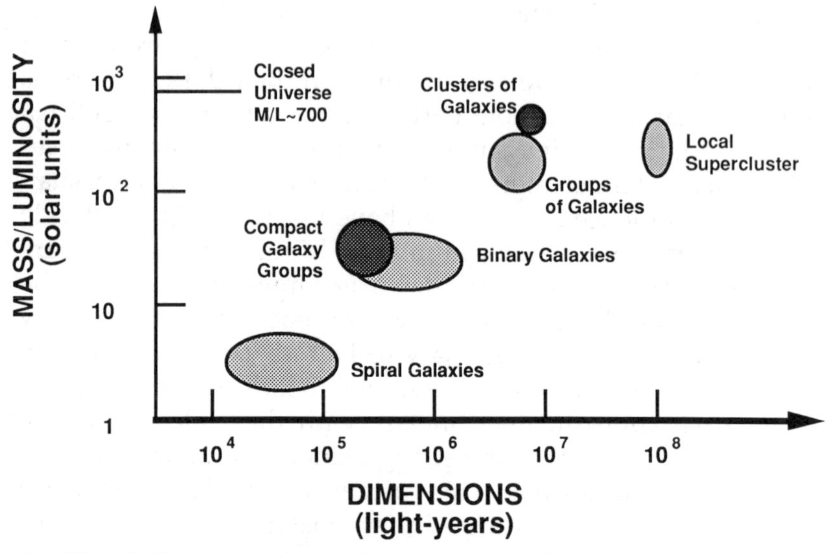

Fig. IX-5.. Ratio of mass to luminosity for aggregates of various sizes. The bigger the aggregation, the closer its ratio of mass/luminosity approaches the value that separates open from closed universes.

the temperature of the surface of the sun. The sky will then be everywhere as bright as the sun's surface is today. Atoms will begin disintegrating at these temperatures into ions and free electrons.

As the density of matter continues to rise, gravity will accelerate the universe's collapse. Only a few hundred thousand years after the sky begins blazing like the sun, it will resemble the center of a star; temperatures will have reached millions of degrees. When they attain billions, in only a few more weeks, nuclei will begin disintegrating into protons and neutrons. A few *minutes* later radiation at trillions of degrees will tear the nucleons apart. Only a ten-thousandth (10^{-4}) of a second later, the thermometer will go off the scale at a hundred million trillion trillion (10^{32}) degrees. Every volume the size of an atomic nucleus today will then contain as much matter as is now in the sun. The density will be 10^{93} times greater than that of water. The known laws of physics fail at these extremes, which may be reached some hundred billion (10^{11}) years from now, so subsequent events are inscrutable to us. Virtually all the information content of the universe will get erased in its rush to crush itself. Life has not the slightest flicker of hope. Even the eternal optimist hoping for a rebound from the universe's navel might have to contend with the reality that the average disorder of the universe increases with each cycle. Eventually, it seems, even a pulsating universe must run down.

Fortunately the evidence presented earlier in this chapter hints that the universe is open and unbounded, and will therefore last forever. Life could then enjoy a future having equally unlimited potential. Intelligence exercising conscious selection will have to be added to evolution via natural selection to guide life's destiny, however, since today's lifestyles will be inappropriate for the changes that await them. At least there will be changes—events continuing to happen—for as far as we can foresee them; that is, in fact, the reason for remaining optimistic about life's immortality.

As mentioned, the sun will expire about five billion (5×10^9) years from now. Presumably our descendants will have learned by then to hop from our aging star to a young one or a long-lived one, prolonging the possibility of life thriving on radiant energy. But this game of "galactic musical chairs" has its limits, too, for after some hundred trillion (10^{14}) years, *all* stars will be dead. Their "remains" will be white dwarfs, neutron stars, or black holes. Black holes may prove to be convenient energy sources for organisms intelligent enough to exploit them. This they may be able to do (in principle, one should add) by shooting matter along very special trajectories near rotating black holes. Although some of the matter will be sacrificed to the hole, part of it can be "whipped" to higher energies by the immense gravitational pull of the black hole. Another way organisms may be able to utilize black holes is by inducing them to collide and coalesce. Black holes that do so can yield a significant fraction of their combined mass in the form of energy.

One should not presuppose that the necessary feats of astro-engineering can never be accomplished, even though we have no idea how to do them at present. One need only recall that our technology currently is only in its infancy.

Some black holes will coalesce on their own, without the help of any intelligent beings. Those in galaxies, for example, will spiral toward the galactic centers because of the "friction" established by their encounters with other matter. Still other star-star encounters will result in quick pirouettes that fling one of the participants from the galaxy. This, to verify its reality, is exactly the use to which Jupiter's gravity was put to "slingshot" the Voyager probes on toward Saturn. Some ten billion billion (10^{19}) years hence, the universe will be a very dilute sea of background radiation punctuated occasionally by some stellar black holes, by some small, cool stars, and, here and there, by

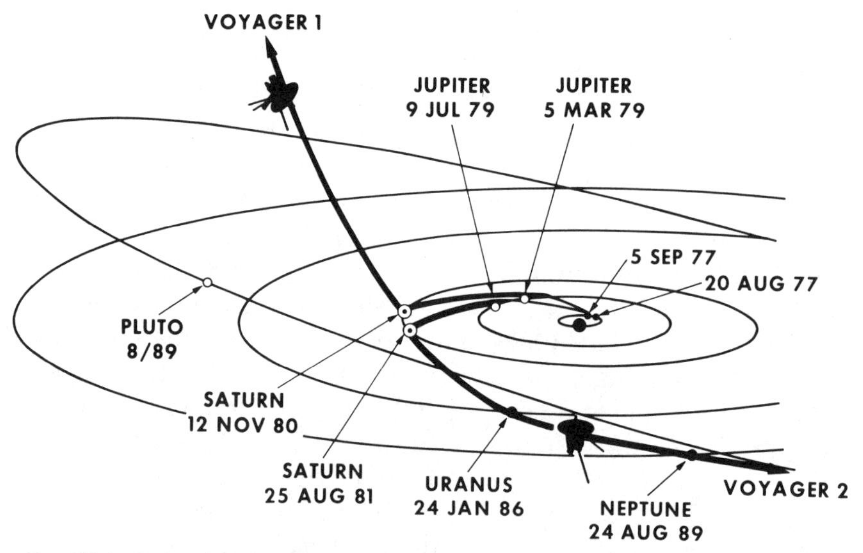

FIG. IX-6. Paths of the Voyager Spacecrafts (Figure VII-8). The giant planets, Jupiter and Saturn, exerted gravitational forces that sharply bent the spacecrafts' trajectories. These interactions were engineered by humans, but in the distant future random encounters between astronomical objects could change the orbits of both. (NASA/JPL.)

a galactic-mass black hole. These objects will persist for an extremely long time; the only noticeable change during this lull will be continuing increase in their separations. Obviously, any life requiring a fixed input of energy will be long since extinct. As energy flows become ever more feeble and the events that produce them ever less frequent, life, if it is to survive, will have to slow its metabolism. Perhaps some clever organisms will learn to do so, or some exceptional organisms will evolve along an appropriate new avenue. The local decreases in entropy that life then will manufacture (at the expense of greater entropy exported elsewhere) must occur in smaller increments separated by longer intervals. Hibernation on an astronomical Rip-van-Winkle scale will become a possible strategy.

To be sure, freezing due to the complete loss of energy generation is still a *long* way off in the future. For example, after 10,000,000,000,000,000,000,000,000,000,000 (10^{31}) years, all protons and neutrons may have decayed into radiation, electrons, and positrons.[2] The particles produced by those decays will be born with some spread in velocities. Like cars on a freeway, they will sort themselves into groups of different speeds. Fast ones will speed ahead of their neighbors until they catch up with others; slow ones will do the same by falling behind. All, meanwhile, will partake in the universe's overall expansion. But their jockeying around represents a flow of matter from a state of thermal agitation (particles spread over many velocities) to one of isothermality (all in one bunch having the same velocity). This is the material analog of the flow of heat energy from hot to cold. Perhaps in some unspecified way life will be able to profit from this flux of particles.

The next major fireworks injecting some energy into the universe will occur at about 10^{65} years. By then black holes of solar mass, which have been evaporating slowly throughout their enormous lifetimes, will begin disappearing explosively and instantaneously. The galactic-mass black holes will do the same

[2] Experiments are in progress to determine the lifetime of the proton.

at 10^{90} years. By 10^{100} years, the few black holes created when galactic black holes collided will have winked out of existence also. Much, much later, at about 10^{1500} years—a number so large its zeroes would fill a typewritten page—all matter not otherwise disposed of will have collapsed into nuclei of iron, nature's most stable element. After what seems forever, these iron spheres will be transformed into miniature black holes, which *eventually* will evaporate. All, in the end, will be turned into radiation.

To understate the obvious, life will not be easy in an infinite universe. It will certainly be different. But neither will it be doomed. The question is, will it be worth living? Freeman Dyson, a pioneer in taking the *long* look at the future, points out that there is little advantage to immortality if it is achieved with only a finite memory, for history has no end. What is the gain if life has to erase knowledge of its origin to make room for retention of newer experiences? Growth of knowledge is equivalent to an increase of order or to a decrease of entropy. Either requires a flow of energy. As energy flows get scarcer

TABLE IX-1. The Future in an Infinite Universe

Epoch	Event
0	Big Bang
10×10^9 years	Birth of solar system
15×10^9 years	Now
20×10^9 years	Sun expires
10^{14} years	All stars have expired
10^{31} years	Protons and neutrons decay into fundamental particles and radiation.
10^{65} years	Stellar-mass black holes explode
10^{90} years	Galactic-mass black holes explode
10^{100} years	Cluster-mass black holes explode
10^{1500} years	All remaining matter nuclei of iron
$10^{????}$ years	Iron nuclei collapse into black holes

and weaker in the far-distant future, order and the growth of knowledge it permits can only increment in very small (and continuously decreasing) steps. But here life might engineer a tradeoff: what is lost in energy can be compensated for in time. An infinite reservoir of time, after all, is available in an open universe. The trick will be to slow thought processes—those resulting in increased knowledge—or even to think them only intermittently. Perhaps in this way life in the future can think an infinite number of thoughts and live forever on only a finite supply of energy. The prospect should encourage us to try. Forever gives us a long time to experiment with different strategies.

Let us ponder one final wild speculation. Many features of the universe seem tailor-made to assure the eventual emergence of life within it. The subject of this chapter—the close balance between matter's gravitation and the momentum of its expansion—is illustrative but not unique. Perhaps some unknown principle operates to guarantee a living universe. If this is so, what of the universe's ability to conserve certain properties? Its total mass-energy is a constant. Matter and radiation may interchange, and matter may assume various forms, but the sum of the two never changes. Could the same be true of life? It could appear in many varieties. With a future looming for at least another hundred billion (10^{11}) years and perhaps an infinite number more, the spectrum from primal slime to *Homo sapiens*, which blossomed in just ten billion (10^{10}) years, is sure to be only a tiny fraction of all forms that will arise. Will life, though, regardless of its manifestation, be conserved? In other words, since the universe took such care to establish a nursery for life, perhaps it will also act to continue life forever. Is life one of the features, like space and time, without which a universe is nonexistent?

For Further Reading

Blitz, Leo, Fich, Michel, and Kulkarni, Shrinivas. 1983. The new Milky Way. *Science* 220: 1233–1240.

Davies, Paul. 1978. *The Runaway Universe*, pp. 159–178. New York: Penguin Books.

Dyson, Freeman J. 1979. Time without end: physics and biology in an open universe. *Reviews of Modern Physics* 51: 447–460.

Dyson, Freeman J. 1988. *Infinite in All Directions*, pp. 97–121. New York: Harper & Row, Publishers.

Frautschi, Steven. 1982. Entropy in an expanding universe. *Science* 217: 593–599.

Krauss, Lawrence M. 1986 December. Dark matter in the universe. *Scientific American* 255(6): 58–68.

Page, Don N., and McKee, M. Randall. 1983 Jaunary-February. The future of the universe. *Mercury* 12(1): 17–23.

Rubin, Vera C. 1983 June. Dark matter in spiral galaxies. *Scientific American* 248(6): 96–108.

Schneider, Stephen E., and Terzian, Yervant. 1984 November-December. Between the galaxies. *American Scientist* 72(6): 574–581.

Seielstad, George A. 1983 *Cosmic Ecology: The View from the Outside In*. Berkeley: The University of California Press.

Trefil, James S. 1983 June. How the Universe will end. *Smithsonian* 14(3): 72–83.

TEN

Biological Copernicanism

It goes against Nature, in a large field only one shaft of wheat to grow, and in an infinite Universe to have only one living world.

Metrodorus

A BASIC PREMISE of this book is that life is an integral component of the universe. Cosmology and biology are not separate disciplines, since life cannot be understood without tracing the origin and evolution of the universe; nor can the universe be comprehended without considering the life residing within it. In particular, life is the only known vehicle for achieving comprehension. These statements are no more than extrapolations from our terrestrial experience with ecology, which has begun to wrap the environment and its living inhabitants into inseparable units called ecosystems.

We are certain, of course, of the existence of only one sample of life in the universe, the one of which we are part. If, however, we believe that sample to be a natural outcome of the operation of the laws of physics, chemistry, and biology, and, furthermore, that these laws apply everywhere in the universe (an assumption we have made implicitly throughout this discourse), then we are led to the view that life may be a common feature of the universe. The philosophical framework within which thought about this question takes place also biases us toward this conclusion.

Consider, for example, the major transformation in thinking that accompanied Copernicus's removal of the earth from the center of the cosmos. Ever since, astronomers and others have been deeply suspicious of any theory that makes our location or epoch in any way special. Moreover, as knowledge of the universe has accumulated, this Copernican outlook has been verified and extended: first, of course, when earth was seen to have companion planets also revolving around the sun, but then later as the sun was realized to be an ordinary star in an ordinary galaxy in an ordinary cluster in an ordinary supercluster.

Darwin strengthened the notion of mediocrity by adding a biological dimension to the physical ones. He pointed out that humans are not so different from other animals, that, in particular, their ancestries have much in common. Subsequent discoveries of molecular biology confirmed the fundamental similarity of structure shared by all earth's living inhabitants—plants and microbes as well as animals. Even the much vaunted human intelligence may not be so special. Chimpanzees and dolphins have been trained to comprehend "sentences" of several words. Here we are carefully avoiding any primate chauvinism, since cetaceans generally seem as curious, communicative, and eager to learn—in short, as intelligent—as primates. Can these two cleanly separated orders be illustrating convergence, the evolutionary arrival at similar attributes via different pathways? If so, then intelligence may be an attribute that is fairly easy to acquire. It may in that case be a fairly common feature of life wherever it exists. The human intellect, then, may be merely typical.

Philip Morrison has focused on this question of convergence toward intelligence, using human speech as a tracer. Speech has without a doubt been a major force driving the advance of collective human intelligence. Each individual has been able to share the wisdom of others by communicating with them. But could forms of communication other than speech have permitted person-to-person exchanges? The evidence from sign languages invented by the deaf indicates that they could have. Sign languages employ nuances of expression—slang, poetry, metaphors, clichés, even a version of music—as complex and as flexible as those in spoken languages. One should also note that similar rules of grammar seem to have evolved among geographically isolated deaf people. So it seems that the principal method by which humans teach and learn—namely, by communication—was not exclusively dependent on possession of specialized organs of speech. The mere existence of several alternate pathways leading to the same result undermines any notion of uniqueness.

Other philosophical underpinnings favor the mundane more than the unique. When expanding knowledge into a previously unexplored realm—and little is as unexplored as the existence of life beyond earth—sound scientific practice assumes a state of quasi-equilibrium. That is, experience tells us that the universe seems much the same year after year and perhaps galactic revolution after galactic revolution (cycles measured in hundreds of millions of years). Experience also tells us that most natural phenomena with any lasting significance arise as small perturbations on a more general stasis. Evolution is much more common in nature than revolution. Now if sharp discontinuities are rare, it is *highly* improbable that the moment at which an investigation began would coincide precisely with one of them. If we biota on earth are unique, however, then "Now" is a moment of sharp discontinuity between an equilibrium state in which the universe was virtually devoid of life and one in which life will be profuse—at least if we follow our first steps into space with eventual colonization. Even if we choose only to communicate via electromagnetic radiation, the Now Epoch is a milestone separating quiet space from purposely noisy space.

None of these arguments is irrefutable; quite the contrary. But our single, certain sample of a living superorganism gives additional cause for optimism about the prospects of likenesses elsewhere. For life arose on earth early in its history. Fossil remains as old as 3.5 billion years (on a planet 4.6 billion years old) have been discovered. The fossils are not of the earliest organisms either, because they are morphologically advanced and represent diverse microbial life. If life arose early here and radiated quickly into diversity, doing so could not have been extremely difficult. Therefore, the argument goes, it could have happened elsewhere.

The diversity evident in these ancient microbial communities provides another argument against the exclusiveness that an earth-populated-only universe would imply, namely, that *diversity will prevail unless some mechanism enforces conformity*. Differences abound on our planet, and the same is true elsewhere. Mars's topography shows peaks and valleys, poles and "temperate" zones. Its canyons display evidence of a similar het-

FIG. X-1. *The surface of Mars is not uniform. High peaks (three at left), deep canyons (running horizontally at center), and polar ice caps (upper left), among other details, illustrate the diversity of environments, past and present. (U.S. Geological Survey, Flagstaff, AZ.)*

erogeneity in time: since the exposed rocks have different colors, their deposition evidently occurred at different times under different climates. Venus, too, has a great plateau whose rock record surely contains traces of earlier different climates. Differences among objects in the solar system are just as striking. Lightning bolting through Jupiter's atmosphere might produce complex organic molecules, just as electric discharges are known to do in chemical mixtures prepared in laboratories on earth. Jupiter's moon, Io, has chemically enriched gases in its atmosphere and hot springs on its surface that are unlike those on other objects in the sun's vicinity. Titan, a moon of Saturn,

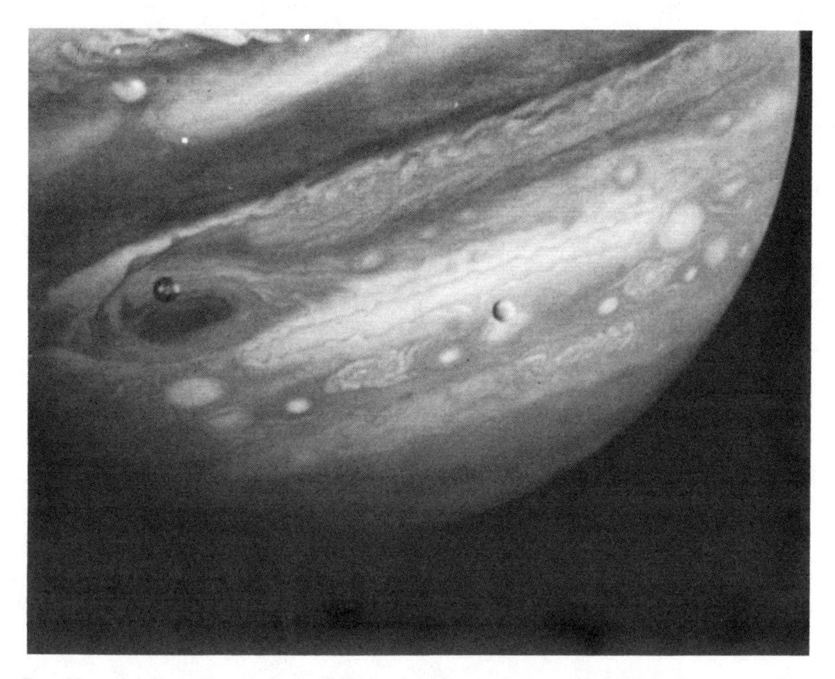

Fig. X-2. Jupiter, its moons Io (left) and Europa (right). No two objects in the solar system are identical. Jupiter's gaseous nature differs from the rocky surfaces of Mercury, Venus, Earth, and Mars. Io, unlike Europa, has active volcanos and oceans of sulfur. (NASA/JPL.)

is blanketed with layers of organic molecules. In fact, no two solar system objects that have been carefully scrutinized are identical. Nor should we expect to find duplicates when we extend our explorations farther. Even well beyond the solar system, in interstellar space itself, the chemistry of molecular clouds is rich and diverse.

This fundamental diversity is central to our consideration of potential extraterrestrial life for two reasons. First, it emphasizes that our previous searches for life, none of which have detected any, of course, have been so limited that their conclusions have no general applicability. The usual qualifier for every search is for life "as we know it." Our preconceptions about the limits within which life can survive, however, are prejudiced by our own experience. Second, the greater the number of different environments, the greater the number of settings in which life can reside. If all environments were identical, the absence of life in one would signal its absence in all. Instead, the possibilities are unlimited. Probably many cannot be excluded as too extreme to permit life's existence.

The two reasons stated in the previous paragraph why diversity favors life throughout the universe are reinforced by recent discoveries of previously unrecognized lifeforms right under our noses, so to speak, in "unlikely" environments here on earth. Consider for starters worms five feet long with no mouths, guts, or openings for waste elimination; or lichen in rock fissures where water and food are scarce and temperatures plunge to −60° Celsius (−76°F); or microscopic fungi on rock surfaces exposed to searing heat, raking winds, and parching humidity. These three examples resemble creatures and conditions we might fantasize to exist on distant planets, but they are in fact residents of our own in niches previously regarded as hostile to life.

The most spectacular of these newly found niches are those around hot springs at the cracks in the sea floor from which tectonic plates diverge. Here, in almost total darkness, at near-crushing pressures and in close proximity to flows of super-

Fig. X-3. Giant tube worm, mussels, and crab at the Rose Garden hydrothermal vent, depth 2500 meters, on the Galapagos Rift. (Courtesy J. F. Grassle, Woods Hole Oceanographic Institution. From Science ***229:*** *715 (1985). Copyright © 1985 by the AAAS.)*

heated water,[1] exist microorganisms. In truth, the presence of bacteria is not astounding, because they are known to be extremely versatile and to exist at many temperatures by exploiting different energy sources. What was completely unexpected,

[1] The temperature of water exiting hot vent fields has been measured to be 270–380°C (~ 520–720°F). Water can be heated beyond its normal boiling point (100°C or 212°F) if under sufficient pressure.

though, was a diversified population of marine animals and the unusual food chain to which they belonged. The foundation of the food pyramid was not direct sunlight—almost none penetrates to ocean depths of a few thousand meters. Instead it was the fixation of carbon by bacteria using the energy of sulfur compounds and other inorganics. Thus, we were excessively conservative in guessing not only where life could exist but also how it could utilize energy to create order.

The energy cycle sustaining life at undersea vents begins when the heat of radioactivity melts rocks in the mantle, which then ooze up in the form of magma at midocean rifts. Seawater percolates down through this newly forming crust, undergoing chemical reactions as it does so. Specifically, sulfates in the water react with iron in the rock to produce hydrogen sulfide (H_2S) and iron oxides. The seawater also dissolves sulfide minerals in the rocks, so that, when it wells up again with the magma, it is rich in these compounds. So rich, in fact, is the sulfide concentration of the heated, recycled seawater that it would be toxic to most organisms, but not to some especially hardy bacteria. They absorb the sulfides, as well as oxygen of photosynthetic origin carried down from the surface waters, and combine the two to manufacture sulfates.[2] Their conversion of sulfides to sulfates releases energy, which the bacteria use to manufacture hydrocarbons. These mats of microorganisms, in turn, are food for a community of invertebrates—clams, mussels, tube worms, and the like. The worms, by letting food-manufacturing microbes reside within them, survive without ingestive and digestive organs of their own. Of course, the microbes, too, gain from substituting for the usual internal organs. They acquire a controlled environment, the need for which arises because the output of seafloor vents fluctuates just as, more familiarly, do those of surface vent fields—the geysers of Yellowstone National Park, for example. At times, the con-

[2] Note that life at the vents is not completely independent from solar energy. The solar input, though, is acquired indirectly.

centration of upwelling hydrogen sulfide is truly lethal, even to the bacteria; at other times, it is too dilute to provide the raw material for energy conversion. If the bacteria live inside clams and worms, they inhabit an environment where these fluctuations are dampened. The success of these symbioses permits other organisms to join the food pyramid. Crabs and other scavengers eat the mussels and tube worms, for example.

These incredible new-found oases alert us to the possibility that life may exist in places, whether on earth or elsewhere, previously considered sterile. The science-fiction flavor of this entire discussion is inescapable. In this instance, however, truth preceded (and perhaps exceeded) fiction. Two of the explorers of deep-sea hydrothermal vents were sufficiently surprised to realize this that they commented:[3]

> The chemosynthetic existence of organisms in the deep sea also suggests a possible occurrence of similar life forms in other planetary settings where water may be present only in the absence of light. It is surprising that, as far as we know, science fiction writers did not turn their attention to geochemically supported complex forms of life until such forms were actually discovered in the deep sea.

The key ingredient seems to be liquid water. Sunlight, or in the general case stellar energy, appears to be less crucial than once believed. There appears no way, though, to achieve complete independence of life from the cosmos. Recall that the heat that drives both the geologic and biologic cycles at earth's undersea vents derives ultimately from the radioactive decay of such elements as uranium, thorium, and potassium buried within the earth. And these heavy elements were synthesized at the moments of explosions by supernovae. So cosmic explosions must precede the formation of stellar systems supporting

[3] Jannasch and Mottl (1985), p. 725.

life (cf. Chapter VIII). The life-sustaining energy sources therefore have cosmic origins; the energy is just not utilized in "real time", as would be the case if the flow of a star's radiant energy were tapped.

Before considering other "unlikely" biocommunities, let us permit ourselves the license to speculate. The interaction between organic and inorganic matter in the superorganism called earth may be drawn tighter by the recent discoveries of thriving biocommunities at deep-sea vents. Bacteria there release abundant gaseous waste byproducts of their chemical syntheses. Among them are methane, hydrogen, and carbon monoxide. The numbers of bacteria at the seafloor cracks and their high metabolism ensure a sizable quantity of gas—and therefore a sizable pressure. Is it coincidence that the pressure is created precisely where plates diverge, or do living organisms assist in the circulation of the crustal plates across earth's surface? If the latter, man's puny engineering feats shrink to insignificance by comparison.

The deep sea is not the only place where life has recently been found despite a suspicion that to look for it there would be foolhardy. Only slightly less bizarre are the microscopic ecosystems recently found in the frigid, dry desert valleys of Antarctica. The analogy to a Martian environment is striking. Temperatures in the Antarctic summer fluctuate between $-15°$ and $0°$ Celsius, and in winter they plunge below $-60°C$. Snowfall is minimal, and in the particular valleys studied, winds descending from an ice plateau sweep most water vapor from the atmosphere. As befits the bleakness of these surroundings, no signs of life are visible *on the surface*. Nor were any seen on the Martian surface. The absence of visible surface life, though, is probably not due only to aridity and cold, since organisms have been found in colder climates and in drier deserts. Rather, the killer is probably the *rapid variability* of conditions caused by gusty winds. Surviving organisms must be able to switch their metabolic activities on and off rapidly in response to these changes. Some—to illustrate these extremes in lifestyles—me-

Fig. X-4. Sandstone cliffs in the high (elevation 1650 meters), dry (note absence of snow) University Valley of Antarctica. Microorganisms reside in the porous rock. In fact, some layers are white because resident microorganisms have leached away the iron compounds that might color them. (Photo: E. Imre Friedmann. From Science *215: 1046 (1982). Copyright © 1982 by the AAAS.)*

tabolize for an annual total of just hundreds of hours. Only during that small accumulated fraction of a year does insolation provide the necessary warmth and melting snow the necessary water to permit activity—and then only in moderating, microniches beneath the exposed surfaces of porous rocks. The rocks' hidden fissures and cavities are successfully occupied by lichens and blue-green bacteria, neither of which appear to be primitive organisms. Rather, both are probably descendants of other organisms forced by environmental changes (for example, glaciation) to retreat into porous rocks. They have had a lengthy history of colonization, since trace fossils discovered in Antarctic rocks have ages approaching four million years. Their significance is best explained by their discoverers:[4]

[4] Friedmann and Weed (1987), p. 705.

> Trace fossils of endolithic microbial colonization in a cold desert environment inspire speculations about possible scenarios for exobiology. Evidence for . . . water during the early history of Mars raises the possibility of . . . primitive life forms there. If such forms were present, . . . these organisms may have withdrawn into porous rocks—the last habitable niche in a deteriorating environment. . . . The search for such structures [near-surface fossils] is a legitimate goal for the future exploration of Mars.

Lichens are symbioses between algae and fungi. Their existence in Antarctica perhaps makes less surprising the discovery of microcolonial fungi on desert rocks. Nevertheless, these latter escaped discovery until 1982. Then they were found in abundance at many sites and in dense populations. Rocks from the Mojave and Sonoran Deserts of North America, the Simpson and Great Victoria Deserts of Australia, and the Gobi Desert of China have surface microcolonies whose typical density is some two hundred colonies per square centimeter (about the area of a postage stamp), but whose peak concentrations approach ten-thousand per square centimeter.

How many other cryptic organisms reside on this planet? Who can say, since many environmental niches remain unexplored. Again, the fundamental reason they have not been explored is the anthropocentric assumption that environments inhospitable to humans must likewise be inhospitable to all other organisms. Extraterrestrial biological investigations can begin in earnest only when we erase these vestiges of human or mammalian chauvinism. Then imagine the richness of possibilities elsewhere.

Just how many "elsewheres" are there? Specifically, how many planets orbit stars other than the sun? The number detected directly is precisely zero, but new observational techniques are beginning to hint at the existence of many. As a matter of fact, today's high probability of extrasolar planets confirms the optimism of the pioneers who, only a few decades ago, called for searches for extraterrestrial intelligence. If they were correct in assuming (in the face of healthy skepticism) ubiquitous worlds where life could reside, might their other speculations not be right also?

The first evidence that stars other than the sun may have planetary systems is that these stars have anomalously low spin (technically called angular momentum). That is, every reasonable calculation of the rotation a star is born with when it collapses from an interstellar cloud predicts much higher rates than are observed. In the solar system, the "missing angular momentum" has been transferred to the planets. As a result, the object with 99.9 percent of the mass of the system, namely, the sun, carries less than one percent of the angular momentum.[5] Perhaps the slow rotation of other sunlike stars is an indication that they, too, are surrounded by planets whose orbital motions contain most of the system's spin.

We previously mentioned briefly that bipolar energetic outflows from young stars suggested the presence of disks encircling them (cf. Chapter VIII). A natural interpretation of such disks is that they are protoplanetary configurations, either detritus left over after plants have formed or material about to condense into planets. All the planets in the solar system, after all, come so close to lying in a single, thin plane that they most certainly must have condensed from a disk, not from a spherical shell of matter. Lately additional evidence for protoplanetary disks is beginning to accumulate. The Infrared Astronomical Satellite (IRAS), for instance, has detected unusual infrared emanations from many stars. Among them is the fifth brightest star, Vega (Alpha Lyrae). As judged from its optical radiation, Vega's surface temperature appears to be almost 10,000°K (17,000°F), but radiation from the longer infrared wavelengths

[5] Angular momentum is the product of mass, velocity, and distance from center of rotation. In the solar system, the masses of the planets are small compared with the sun's, but the distances and velocities are large (cf. Figure IX-3).

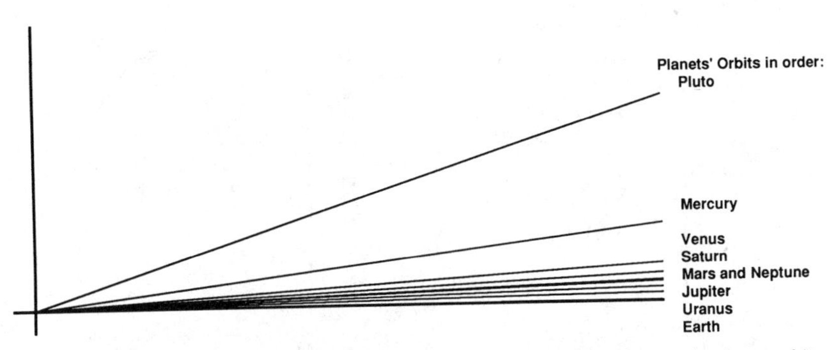

Fig. X-5. Edge view of the planes of solar system orbits. The planets in these orbits show evidence of condensation from a disk, not from a spherical cloud.

exceeds the amount expected for a thermal body at this temperature. At a wavelength of 25 microns (0.001 inches), the excess is about thirty percent, whereas at 60 and 100 microns, the excess is much greater—by factors of 10 and 20, respectively. Furthermore, the infrared radiation emanates not from a single point, as from a star, but rather from an extended region whose physical dimension is about 80 times the distance between sun and earth. The interpretation is that Vega is surrounded by a ring of dust particles whose extent is comparable to that of the solar system. The dark dust grains absorb heat radiated by the hot central star. This warms them to about 90° Kelvin (−300°F), at which temperature they reradiate their energy at the much longer infrared wavelengths. The dark cloud contains about a thousandth (0.001) of a solar mass, which happens to be the approximate summed mass of all the planets and asteroids in the solar system. The sizes of the dust particles are poorly determined. In particular, the presence of objects of planetary mass cannot be excluded.

Once infrared astronomers called attention to these protoplanetary disks, optical astronomers were quick to supply photographic evidence. By blocking the glaring light from the star β Pictoris, to pick a specific example, they were able to see edge on a flat disk of orbiting material. The disk extends to four-

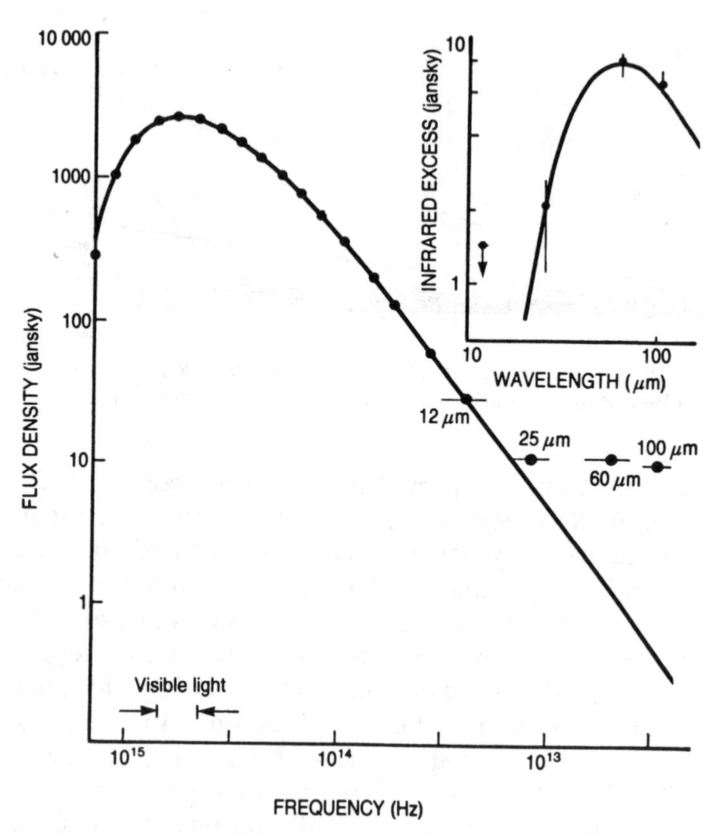

FIG. X-6. Optical and Infrared Radiation from the Star Vega. The curve is expected for radiation from a blackbody whose temperature is 9700 K. At the lowest frequencies a second "hump" (inset) indicates radiation from an 85 K blackbody. The latter hump could be re-radiation by dust particles encircling the central star in a disk and heated by it. (From Physics Today *37 (5): 17(1984).)*

hundred times the earth-sun separation (called an astronomical unit or AU) on either side of the star. It is only 50 A.U. thick at a radius of 300 AU, so the particulate matter lies within 5° of the plane of the system. Again, one cannot be sure if the material seen is what remains after the formation of planetary companions to β Pic or is what will be used in their formation.

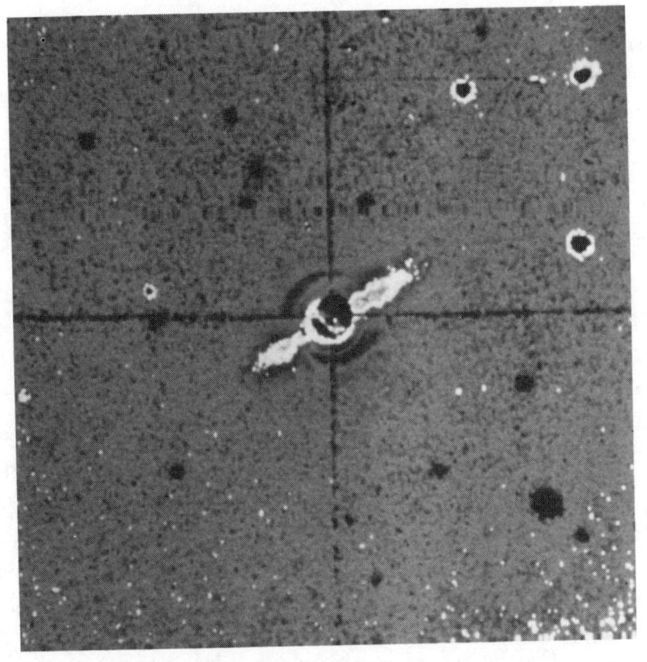

FIG. X-7. A circumstellar disk, around the star β Pictoris (whose light is blocked out). (NASA.)

In either case, and considering that β Pictoris is rather ordinary, one is driven to regard planets in other stellar systems as likely and frequent occurrences.

Some scientists have even suggested the possibility of direct detection of planetary size companions to stars. Most of these observations are also made at infrared wavelengths, where the contrast between bright star and dark companion is considerably reduced. One of the best examples is a low-luminosity infrared source detected near T Tauri, the prototype of an infant star. One interpretation suggests that the infrared luminosity is heat from a very large condensing protoplanet. This interpretation is, however, controversial. Nevertheless, evidence for similar companions in other stellar systems is accumulating via a new observational technique (called infrared speckle interferometry).

Sometimes a small companion that remains truly invisible does, nevertheless, reveal its presence. Even though the companion is dark relative to its dominant partner, it exerts a gravitational pull on its parent star, an object that can be seen. The result is the slightest wobble in that star's position (cf. Figure II–1). Barnard's star is the best studied case, but a handful of others reveal perturbations suggestive of an unseen partner. The difficulty is that the measurements strain the limits of existing instrumentation and are therefore the frequent subject of dispute.

The dance two objects in gravitational embrace undergo displays itself in another way besides position wobble. Specifically, the narrow features in the spectrum of a star revolving around the center of mass determined by it and its companions should shift back and forth in a periodic fashion. These features, called spectral lines, come from transitions between discrete energy levels of the various atoms in the star's atmosphere. Since the energy jumps are quantized, the frequencies of the corresponding radiation are narrowband; hence the name, lines. Because of the Doppler effect, when the star appears to be going away from us, its spectral lines should be

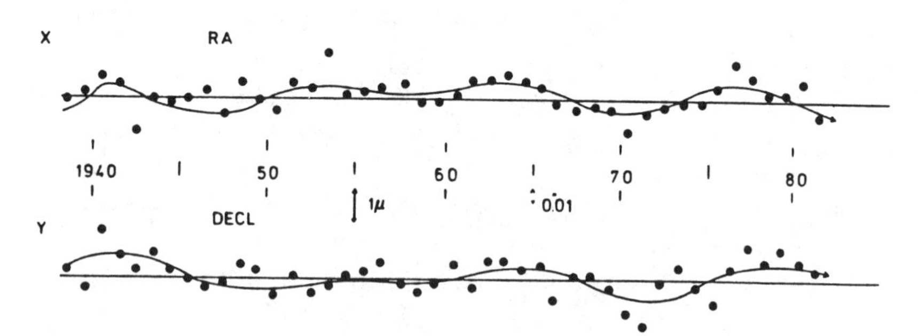

FIG. X-8. Wobble in the position (both angular coordinates) of Barnard's star. Curves show wobbles expected if star has two planetary companions. (Courtesy P. van de Kamp.)

shifted toward the red; when approaching, the shift should be toward the blue; when cutting across the line of sight twice each orbital revolution, no shift at all should be seen. As in the case of wobble, these spectroscopic measurements are difficult and must accumulate for several years.[6] But a careful six-year survey of sixteen nearby sunlike stars has revealed strong indications of companions around two of them and hints of companions around five others. All the companions have masses in the range of one to ten times Jupiter's mass (itself 318 times earth's mass and ~ 1/1000 times the sun's mass). Whether this tentatively advanced evidence will hold up remains to be seen. But the preponderance of all the various measurements points in the same direction, toward the existence of extrasolar planets.

Too many observations and theories point to the existence of planets accompanying stars to suggest anything but their common occurrence. The questions of which are habitable and indeed whether some are already inhabited are much farther from being answered. Our biological ignorance is too profound to permit definitiveness. Nevertheless, our terrestrial experience warns against excessive caution regarding possibilities. Life in the one sample available demonstrates resiliency and adaptability whose limits have not been defined. Life also demonstrates a relentless drive to occupy unexploited environmental niches. The most recent manifestation of that drive, namely the effort to colonize space, is now in its earliest stages.

One can argue that, if a drive to colonize space is a universal characteristic of life, older and wiser civilizations than our own could already have embarked upon this effort. Since only a few hundred million years—a short time in a cosmic context—would be needed to colonize the entire Milky Way Galaxy, we

[6] Jupiter, for example, has an orbital period about the sun of approximately twelve years. Comparable planets about other stars should have similar periods. To be certain, astronomers would like to record the changes over several periods.

should by now have been visited, but there is no evidence we have been. A pessimist would conclude there are no other civilizations. An optimist would point out how much easier and cheaper (in terms of energy expended) a contact is if it is made using electromagnetic signals instead of interstellar travel. We—students of the universe, admirers of its wonders, and possessors of its most information-rich structures, the living ones—should pursue our scientific investigations to settle the issue with all the vigor we can muster. But as we do so, we should assume our civilization is the only sample of life in the universe. For that assumption forces us to act in ways calculated to preserve the continuity that led to our presence, but that seems nowhere near final fulfillment. We inherit an obligation to permit life's progression to prosper. To act otherwise is analogous to a species ceasing reproduction because its descendants will have done nothing for their antecedents. The attitude is selfishness in a time dimension, that is, idolatry of one's individual transient existence as contrasted with cognizance of membership in a lifestream both more comprehensive and potentially eternal. If our conservative assumption is wrong and life already exists beyond earth, our actions to conserve and preserve have not been wasted. Earthlife then serves as one manifestation of a much more generalized phenomenon. Loss of even one sample from the universe of the living is akin to extinction of a particular species in the terrestrial superorganism. Neither event is the kind we want to cause to occur more frequently than its natural rate.

If biological Copernicanism is a realistic extrapolation of our ever-widening knowledge of the universe, incredible wonders await our discovery. Each bit of information we acquire brings us closer to stumbling upon them. The open-ended search for understanding of our cosmos promises to be the most worthwhile of all human endeavors. If it eventually reveals that the universe abounds with life, and also reveals ways to communicate with the most advanced representatives of it, our intellects may be instantaneously stretched by an amount that would, if left to our own devices, require millions or billions of years to acquire.

For Further Reading

Awramik, S. M. 1986. New fossil finds in old rocks. *Nature* 319: 446–447.

Brin, Glen David. 1983. The "Great Silence": the controversy concerning extraterrestrial life. *Quarterly Journal of the Royal Astronomical Society* 24: 283–309.

Brownlee, Shannon. 1984 July. Bizarre beats of the abyss. *Discover* 5(7): 71–74.

Eberhart, J. 1983. Vega & Co.: What's being born out there? *Science News* 124: 116.

Edmond, John M., and Von Damme, Karen. 1983 April. Hot springs on the ocean floor. *Scientific American* 248(4): 78–93.

Felbeck, Horst. 1981. Chemoautotrophic potential of the hydrothermal vent tube worm, *Riftia pachyptiba* Jones (Vestimentifera). *Science* 213: 336–338.

Friedmann, E. Imre. 1982. Endolithic microorganisms in the Antarctic cold desert. *Science* 215: 1045–1053.

Friedmann, E. Imre, and Weed, Rebecca. 1987. Microbial trace-fossil formation, biogenesis, and abiotic weathering in the Antarctic cold desert. *Science* 236: 703–705.

Goldsmith, Donald. 1988 February. SETI: The search heats up. *Sky & Telescope* 75(2): 141–143.

Jannasch, Holger W., and Mottl, Michael J. 1985. Geomicrobiology of deep-sea hydrothermal vents. *Science* 229: 717–725.

Morrison, Philip. 1981. Reflections. In *Life in the Universe,* John Billingham, ed. Cambridge, MA: The MIT Press, pp. 421–433.

Schopf, J. William, and Packer, Bonnie M. 1987. Early Archaean (3.3-billion to 3.5-billion-year-old) microfossils from Warrawoona Group, Australia. *Science* 237: 70–73.

Schwarzschild, Bertram M. 1984 May. Infrared evidence for protoplanetary rings around seven stars. *Physics Today* 37(5): 17–20.

Smith, Bradford, A., and Terrile, Richard J. 1984. A circumstellar disk around β Pictoris. *Science* 226: 1421–1424.

Staley, James T.; Palmer, Fred; and Adams, John B. 1982. Microcolonial fungi: common inhabitants on desert rocks? *Science* 215: 1093–1095.

Thomsen, Dietrick E. 1982. Seeking the planets of other stars. *Science News* 121: 424–426.

Trefil, James. 1983 January-February. Alone in the universe. *The Sciences* 23(1): 16–19.

Waldrop, M. Mitchell. 1987. Extrasolar planets, maybe—but brown dwarfs, no. *Science* 236: 1623–1624.

Walgate, Robert. 1983. Emerging solar systems in view. *Nature* 304: 681.

ELEVEN

The Riddle of Being in the Here and Now

He believed in an infinite series of times, in a dizzily growing, ever spreading network of diverging, converging, and parallel times. This web of time—the strands of which approach one another, bifurcate, intersect, or ignore each other through the centuries—embraces every possibility. We do not exist in most of them. In some you exist and not I, while in others I do, and you do not, and in yet others both of us exist.

Jorge Luis Borges

LIFE CANNOT BE CONSIDERED apart from the universe that gave birth to it and that provides an environment for its presence. The universe began in a state of maximum chaos—equivalent to minimum information or maximum entropy—and proceeded to unfold in such a way that, at least locally and temporarily, entropy could decrease. Various structures—galaxies, stars, planets, asteroids, and the like—were manifestations of this process set in action at the big bang itself. So, too, was life a natural development in this progression of localized entropy reduction.[1]

Precisely why the universe would tend to establish pockets of high information content is a question as yet unaddressed. Perhaps the question is meaningless. Maybe the universe just is, and whatever occurs within it just happens. If so, the fact that the universe is governed by the laws of nature that we observe to hold true has no explanation. An "explanation" may exist, however, if acts of observation are prerequisites for the existence of a universe. We rephrase Leibniz's question, "why is there something rather than nothing?", to ask, "is there something if nothing ever notices it?" If the answer is negative, then intelligent life is necessary in the universe, and various past conditions can be deduced from the fact that we had to be here to deduce them. This new kind of effect-to-cause reasoning defines the *anthropic principle.* Carrying this argument to its extreme, one can *require* the universe to behave in such a way as to *guarantee* lifeforms sufficiently intelligent to reconstruct the universe's history. In this form, the statement is called the

[1] The localized reduction of entropy occurs, of course, at the expense of a more than compensatory increase elsewhere.

strong anthropic principle. In this chapter, then, we consider whether the universe can be understood only if life is a necessary component of it.

Let us review how our very existence places constraints on gross features of the universe, for example, why the universe is as ancient and as enormous as it is (some tens of billions of years and tens of billions of light-years, respectively). The answer given by the anthropic principle is that we are here to record its age and size. The reasoning, you may recall, starts with the assumption that the universe, to be real, demands that something have an awareness of it. Awareness requires intelligence, and intelligence requires life. Life is constituted from heavy elements. Heavy elements result from thermonuclear combustion over long timespans, that is, from cooking at temperatures of tens of millions of degrees for tens of billions of years. The universe is therefore observed by us to be old, otherwise we would not be present to note that fact. Likewise, we observe the universe to be immense because the long times necessary to foster our existence produced, by expansion, a universe of just such gigantic proportions.

We have pointed out in earlier chapters that the age, size, and composition of the universe are also dependent upon the fact that the expansion of space is occurring at almost (or perhaps exactly) the rate distinguishing open, infinite universes from closed, finite ones. Had the expansion rate been strongly decelerated, the universe's total lifetime from big bang to ultimate collapse would have been too brief to permit time adequate for the incubation of life. Had the expansion rate instead been negligibly decelerated, matter within the universe would have dispersed before congregating into galaxies, planets, or people. Our presence consequently argues for a space not too deviant from the flat one occurring when the gravity of the universe's total mass-energy precisely balances the momentum of the expansion with which it was launched.

Other initial conditions were likewise crucial if they were to be followed, eventually, by creatures who could reconstruct them. For instance, the background of radiation filling all the space around us has cooled by now to a few degrees Kelvin. If it were just a few hundred times hotter—and what is a puny factor of a hundred or so when we know the temperature has ranged over some thirty orders of magnitude more than that?—the chemical reactions taking place within every living organism would be precluded. The determining factors of this temperature were the initial density of mass and energy and the rate at which expansion diluted these densities. The mass-energy density in turn depended upon the numbers of photons, quanta of radiant energy, per proton; and the numbers of protons surviving the universe's explosive origin demanded a slight excess over anti-protons. The details are less important than the fact that many physical parameters had little latitude if life was to unfold as a consequence of them.

In previous chapters we discussed how the existence of observers even speaks to the magnitudes of the forces of nature—the strong and weak nuclear, the electromagnetic, and the gravitational. A stronger electromagnetic force, we now know, would have pulled electron orbits inside atomic nuclei, so that chemistry and hence life would have been impossible. It would also have blown apart heavy nuclei, those with many protons, unless the strong nuclear force had itself been strengthened. But a stronger strong force would have permitted stable diproton nuclei whose subsequent fusion into helium nuclei would have been so rapid that all the hydrogen resident in the early universe would have been consumed during its first few minutes; the stellar energy sources for powering life could never have arisen. Weakening electromagnetic and nuclear forces would have had similarly disastrous consequences.

Gravity, too, had to be tightly constrained if life were ever to be present. The chain of reasoning leading to this conclusion starts with the observation that the life exemplified by earth's superorganism has prospered only because of the vast thermal disequilibrium permitting a star's energy to flow into the cold space surrounding it. Furthermore, this flow has had to persist over long durations to permit self-organization of a superorganism. Both these conditions necessary for our one sample of life, and therefore for other samples like it, would be affected if gravity had been weaker or stronger. On the one hand, if its pull had been slightly strengthened, all stars would have been blue giants, orders of magnitude more massive than the sun. Such monsters consume fuel so profligately that their lifetimes are too short to provide sufficient incubation time for life. On the other hand, if the force of gravity had been slightly weakened, only red dwarfs a fraction of the sun's mass would have formed. While their lifetimes are immense, their power outputs are small. As a consequence, the zones around them where temperatures are appropriate for (our kind of) life are both narrow and close in. The likelihood that a planet would be found in an appropriate zone is small compared with the case of sunlike stars.

Not only must stars radiate life-sustaining energy; they must also manufacture and distribute the heavy chemical elements from which life is assembled. The distribution is accomplished when envelopes of stars are blown into space by the eruption of supernovae. The explosive force disrupting the stars derives from a rapid buildup of neutrinos beneath their surfaces. The neutrinos are consequences of weak nuclear interactions, so changes in the weak nuclear force likewise endanger life's prospects.

For all the foregoing reasons and for several more as well, the universe we see is, within narrow limits, the only one we could see. That fact alone does not argue that we had to come into existence. There could, one supposes, be sterile universes, but only if the existence of anything is possible when it is never noticed. We might also note that many of the cosmic coincidences cited would seem distinctly less remarkable if life could exist under a broader range of conditions than the ones *we* require—for instance, as a type of plasma made of elementary

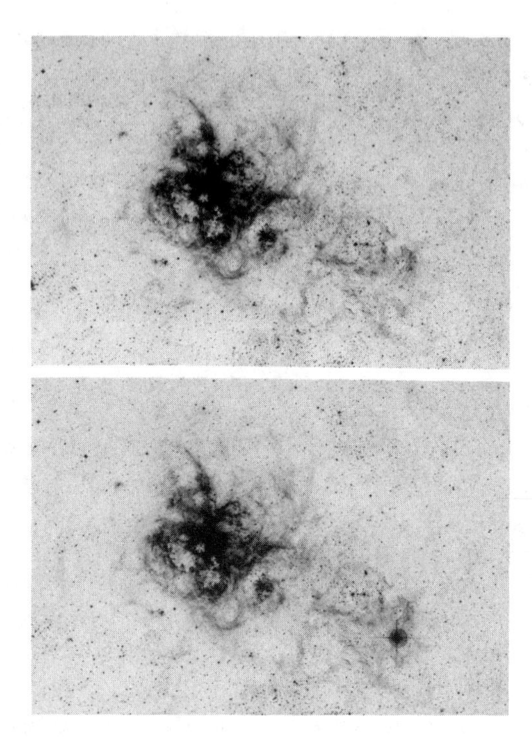

Fig. XI-1. Supernova in Large Magellanic Cloud. Before explosion of February 23, 1987 (top) and after (bottom). Explosion was so violent that the destroyed star (at lower right, bottom) was bright enough to be seen without the aid of a telescope. The heavy chemicals future generations of life in the LMC's Tarantula Nebula will need are being dispersed, and in some cases manufactured, by the explosive force. (Copyright © 1987 Royal Observatory, Edinburgh.)

particles, or in a thermodynamically differentiated environment besides the one on a gradient between hot sources and cold sinks.

An analogy demonstrates how *post hoc* and earthnocentric arguments can mislead us into equating the improbable with the miraculous. Pause to consider how special earth must be to serve as a habitat for life. It must provide an abundant supply of such atoms as hydrogen, carbon, oxygen, nitrogen, phosphorus, and sulfur, admixed with small but vital contributions of calcium, potassium, and sodium. Yet the molecules formed from these atoms must not be poisonous, as are some of the dominant ones on and around other solar system bodies. Temperatures must be such as to permit liquid water, and a stable, enduring supply of free energy must bathe earth's environs. Gravity must be robust enough to anchor an atmosphere, but not so great that organisms cannot move freely or even fall without catastrophic consequences. An ozone layer to shield

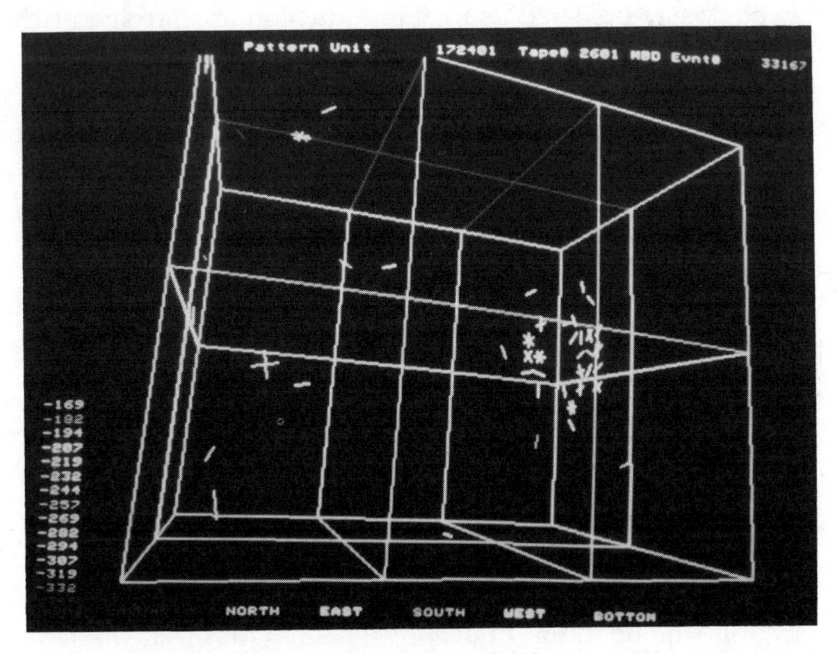

Fig. XI-2. Signature of supernova's neutrinos. The "telescope" is an enormous tank of water located deep in a salt mine. Eight neutrinos—way above the background level in this well-shielded telescope—produced flashes in this tank just three hours before the visible light from the 1987 Supernova in the Large Magellanic Cloud (Figure XI-1) was detected, a coincidence in time that leaves little doubt as to the source of the neutrinos. Not only was a new discipline, neutrino astronomy, opened, but also the operation of the weak nuclear force was confirmed. (Brookhaven National Laboratory.)

ultraviolet radiation and a magnetic field to deflect cosmic subatomic particles are also necessities.

Following Darwin, however, we no longer marvel at why earth is so benevolent to life. Instead, we consider how life has adapted to earth's parameters and perhaps even modified them to improve life's sustainability. In other words, if conditions had been different, life would have packaged itself in different units. We know, for instance, that creatures at seafloor vents can exist on chemical diets, and need not utilize the sun's energy directly to do so. Likewise, a stronger surface gravitation would have bred differently constructed creatures. In addition, life obviously predated its own introduction of an oxygen-rich atmosphere having an ozone layer. So our location in the universe is not random. We exist in the locale where, and at the time when, conditions are ideal for our presence. Conversely, we (meaning all earth's life) are selected to be matched to the conditions prevailing in our location and epoch. Even if the earth had been completely inhospitable, other planets about other stars, perhaps even in other galaxies, might not have been so. Another author of a different biological type could be writing this same message for a different readership.

Can we extrapolate this reasoning to explain why life inhabits our universe? That we consider ourselves (in this case, representatives of all life in the universe) the improbable consequence of several cosmic coincidences implicitly assumes that there may be alternative universes. Otherwise, how can likelihood be assessed? Try to imagine an ensemble of universes, each characterized by a different combination of values for the fundamental constants (the speed of light, the charge of an electron, the quantum of action, the strength of gravitational attraction, and many others). Each universe will therefore be governed by different laws of nature. We can even permit each to evolve from a different initial balance between mass-energy and motion. In the context of such a superspace of multiple universes, the fact that life intelligent enough to include observers exists in (at least) one universe and at the present time

Fig. XI-3. Spiral galaxy Messier 81 or NGC3031. In it are so many stars like our sun that life may have found a suitable habitat near some of them. (National Optical Astronomy Observatories.)

is no longer miraculous. It is instead a manifestation of cosmic natural selection. The logic reduces to a tautology: observers exist only in universes that permit their existence. Finding life in this universe is no more astonishing than finding life dependent on an enduring flow of heat and light in the solar system. In both cases, that is where and when conditions are appropriate. Our living universe is indeed improbable in the sense that an infinite number of other universes are uninhabitable. But we could not be elsewhere.

Is a superspace containing an infinity of universes an absurdity? Where do the realms of space and time hide their existence? At least three possibilities are already extant. In one, the ambiguities inherent in the quantum mechanical world of subatomic particles are erased by permitting all possible outcomes of an event to proceed simultaneously. Recall that probability replaces certainty in quantum phenomena. Particles have certain probabilities of being present at various locations. Detection of a particle at a specific location, however, means it is there, not elsewhere. But physicist Hugh Everett has suggested that the particle actually exists simultaneously at every location its probability permits it to be. Whenever a measurement locates a particle at a definite position, it merely selects one world from among an infinity of simultaneous ones—and all continue to exist, independently, albeit many lack conditions amenable to observers. Throughout the history of matter, every interaction between pairs of particles splits reality into branches representing every possible outcome of the interaction. Our existence is just one particular path through a maze of infinite branches occurring infinitely often. Along that one path, life is *inevitable,* although still rare because nearly all of the coexisting paths are inhospitable.

Another mechanism, mentioned earlier, for creating an ensemble of universes arises from the frothing vacuum state before the big bang erupted. Bubbles of space and time appeared out of the nothingness. In at least one—that which we inhabit—the inflation of the bubble of space-time was so rapid that the flatness, isotropy, and uniformity of space so vital to life's existence resulted naturally. But other bubbles of space-time had come and gone, each vested with some distinguishing peculiarity. Again, life arose in the one in which it was possible.

The ensemble of universes could just as easily be arrayed along the time axis as scattered across space. That is to say, universes could occur consecutively instead of simultaneously. We have already discussed the prospect that our universe may terminate in a collapse to a singularity. In addition, we pointed out how the big bang resembles the final explosive evaporation of a black hole of mass equivalent to a complete universe. So perhaps our universe was preceded by another and will be followed by yet another—on and on through endless cycles (see Figure VIII–10). When all of space and time collapse at the demise of any single cycle, the laws of nature disappear. After all, they have meaning only so long as a frame of reference exists. It is not implausible to imagine that at each reemergence from these eradications new laws of nature and new fundamental constants blossom forth. Then any cycle in which life occurs is one whose initial conditions are precisely tuned to make life an inevitable consequence.

All three of these types of superspace extrapolate the breadth of a Copernican outlook to its limit. Not only are our planet, star, galaxy, cluster, supercluster—and probably superorganism—typical and mundane, but so is the whole universe they reside in. Of course, ours is special in hosting conditions in which life can exist, but so are all other universes special in some way. The conditions we marvel at, because without them we would be absent, would eventually and naturally occur somewhere at sometime. Existence has all of eternity and an infinity of space to demonstrate that this is so. Improbability bordering on uniqueness gives way to inevitability.

Has nature spent forever stirring random ingredients until their mixture was just right to permit the introduction of organisms capable of fathoming nature? Or was this progression toward observership more necessary than arbitrary in ensuring the reality of nature? The answer depends upon whether one believes that only that which is perceived is truly real. If this is the case, then observers are necessary. Their presence rescues a universe from nonentity by recognizing that universe's existence. Again, quantum mechanics hints at a need for observers. They become an integral part of the system they seek to study. When they insert apparatus into a physical system to measure its properties, they change the state of the system. Furthermore, apparatus inserted to measure one particular property excludes the possibility of measuring a complementary property. Either-or examples include location and velocity, energy and time, and several others. So what the observer decides to measure affects what in fact he is able to measure. Indeed, some of the properties of the system only acquire meaning when they are measured. All of this suggests that observers are not spectators external to a physical system but *participants* in determining the state of the system. Even so, perhaps observers need not be present in all physical systems at all times. It may suffice that some exist somewhere and at some time, provided they are sufficiently astute to extrapolate from observations of their own surroundings to the existence of all others. If these observers' minds can determine all that exists in addition to what their own observations reveal, then perhaps all parts of superspace for all eternity acquire reality by virtue of this recognition.

Whether life is an incidental or an essential part of a universe need not affect our behavior. We are here now, and able to ensure our existence indefinitely into the future. Why chance prematurely eliminating the one feature our particular spacetime was uniquely constructed to provide? The horizon of our living universe extends farther in the forward time direction than in the backward. Life therefore is in its merest infancy. Who can view a youngster without envying the world of potential that awaits him? In the same spirit, who can jeopardize a form of entropy reduction whose maturity lies so far in the future? Today is the beginning of a bio-eternity.

For Further Reading

Carr, B. J., and Rees, M. J. 1979. The anthropic principle and the structure of the physical world. *Nature* 278: 605–612.

Davies, P. C. W. 1978 May-June. The tailor-made universe. *The Sciences* 18(3): 6–10.

Davies, Paul. 1980. *Other Worlds*. New York: Simon and Schuster.

Davies, P. C. W. 1982. *The Accidental Universe*, pp. 110–130. Cambridge, England: Cambridge University Press.

Finkbeiner, Ann. 1984 August. A universe in our own image. *Sky & Telescope* 68(2): 106–111.
Gale, George. 1981 December. The anthropic principle. *Scientific American* 245(6): 154–171.
Linde, Andrei. 1987 September. Particle physics and inflationary cosmology. *Physics Today* 40(9): 61–68.
Wheeler, John Archibald. 1977. Genesis and observership. In *Foundational Problems in the Special Sciences*, pp. 3–33. Eds. Butts and Hintikka. Dordrecht, Holland: D. Reidel Publishing Company.
Wheeler, John Archibald. 1981 June. This participatory universe. *Science 81* 2(5): 66–67.

TWELVE

Humanity

It is possible to believe that all the past is but the beginning of a beginning, and that all that is and has been is but the twilight of the dawn. It is possible to believe that all the human mind has ever accomplished is but the dream before the awakening. . . . We are creatures of the twilight. But it is out of our race and lineage that all minds will spring . . . that will reach forward fearlessly to comprehend this future that defeats our eyes. All this world is heavy with the promise of greater things. . . .

H. G. Wells

Names foul in the mouthing.
The human race is bound to defile, I've often noticed it,
Whatever they can reach or name, they'd shit on the morning star
If they could reach. . . .

A day will come when the earth will scratch herself and smile and rub off humanity.

Robinson Jeffers

WELLS AND JEFFERS spell out our options, the former radiating optimism, the latter pessimism. Is it significant that the three-quarters of a century separating these eloquent spokesmen coincided with mankind's mastery of completely global powers? Could the resultant changes man has produced in the biosphere account for the transition from hope to despair? The answers cannot be found by consulting our predecessors, neither the pair cited nor any others. The answers lie within ourselves, at least within those of us with awareness of the opportunity offered by entrance into a new era of cosmic evolution. In grasping the significance of this opportunity, both Wells and Jeffers speak the same message. They both comprehend the abilities humans have for creating a future and disagree only as to how responsibly we shall use them. Neither underestimates our potential. Nor should we.

We should not, then, await final resolution of the question raised in the preceding chapter of whether humans (or other intelligent observers) are incidental or essential components of the universe. We should instead take advantage of the fact that the universe has offered us a chance to define *our own* significance. We are in the right place at the right time to be creatures of consequence. In the nearly four billion years life has been present on this planet, no other creatures have been similarly blessed—or cursed. We can cease making gods in our own images and ourselves assume some of the tasks we assign them. Now is not the time to focus on which past conditions and events enabled our existence, as much as it is a time to contemplate—even to *select*—which ones will permit life in the future. The arrow of time, of only finite length in the past, may extend forever into the future. Life, as embodied in the superorganism we belong to, need not be present for the entire duration . . . but it could be. It will not be, however, unless we refine the appreciation that time's future arrow has infinite length by unshackling our imaginations. And what bolder way to do so than by freeing ourselves to think about space.

We have initiated a new era by breaking our ties to the home planet that nourished us. The only comparable event in the history of earth's superorganism was when life crept out of the sea and onto the land. But that transition still offered only a finite niche to exploit. Space offers *limitless* possibilities. At this crucial moment of transition, one could say that space's principal virtue is that it is not the earth. Since it is still essentially untouched and unspoiled (at least beyond the burgeoning band of geostationary satellites), it remains full of possibility. By traveling there—*even if only vicariously*—we can abandon the routine of habit, dismiss worn-out ideology, experience the thrill of the new rather than the weariness of the commonplace, and even speculate on the possibility of other-than-carbon-based life. In other words, entrance into the new environment of space offers the opportunity (one time only) for fresh starts. And nothing bounds their range. The opportunity is not only wide open; it can also bring about change even before we have the means to exploit it. To cite one example, just the *knowledge* that humans will reside permanently elsewhere than on earth—never mind how or when—drives people on our planet toward cooperation, because life in the first colonies will be challenging enough without the added burden of competition. Since much of the advantage of life's advance off the earth is therefore mental speculation and extrapolation, we are offered the luxury of *idealism*. To operate every day in the real world requires realism, but to improve upon that world requires idealism. Space thus offers us the chance for improvements.

Space helps free our imaginations not just in terms of physical dimensions, but also with regard to time. The cycles of the superorganism we contribute to and profit from are much longer than any single human lifetime. But it is these long cycles that have cosmic significance, not the infinitesimal periods for which any one of us participates in them. The hundred-million-year domination of earth's biota by dinosaurs, to pick one example, strongly influenced the subsequent evolution of the biosphere; but no individual *Coelophysis*, say, cast a shadow long enough to be noticed. Our focus should therefore be on the entire lifestream, an entity whose coordinates stretch along the time axis, and not on any particular instant occurring within it. We must act as if *continuity* between past and future has significance.

The necessary change in outlook will not be brought about easily. People in western, capitalist societies in particular operate within systems geared to provide *immediate* self-gratification. Incessant advertising reminds each of us that we are living now, that we deserve to live well, that our rewards should come to us immediately. This attitude has dire consequences for our offspring, because some of what we do impoverishes the world for those who will follow. Our actions, too, are governed by the second law of thermodynamics, so that every "improvement" we engineer (measured by an increase in order) results in a greater increase of entropy (disorder) elsewhere. And let us not minimize the magnitude of the disorder we create: humans are today the most important agent modifying the surface of the planet. Not only are our modifications drastic, they also are accomplished ever faster. We are consequently challenging Gaia to adjust more rapidly than she has had to in her preceding five billenia (5×10^9 years).

On the average, twenty tons of mineral material must be excavated annually for each man, woman and child who lives in the modern industrial world.[1] Since there are a billion (10^9) such people, their total annual consumption approximately equals the sum of the masses moved annually by the major natural geological processes of ocean-crust formation, mountain build-

[1] This according to Fyfe (1981), who quotes other sources.

Fig. XII-1. Tons of material are excavated each year for every resident of a developed nation.

ing, and erosion. And this calculation does not even include the amount of matter man moves in agriculture! On top of this, the fossil fuel consumption used to derive energy sufficient to find, move, and process this mineral economy adds to the carbon dioxide reservoir in the atmosphere an additional one percent every year. If this seems small, consider that it translates into a temperature increase of 2° or 3° Celsius by the middle of the next century, one of the largest and most rapid temperature changes in the history of the planet. The buildup of atmospheric carbon dioxide is also accelerating because of the removal of the vegetation that consumes it in the Amazon basin—at a rate that may denude that basin in the next century. If it does, the origin of *one-fifth* of the fresh water flowing annually into the oceans may be adversely affected. We humans are obviously experimenting with the global climate on a grand scale, but the outcome of our experiment is too complex for us to predict.

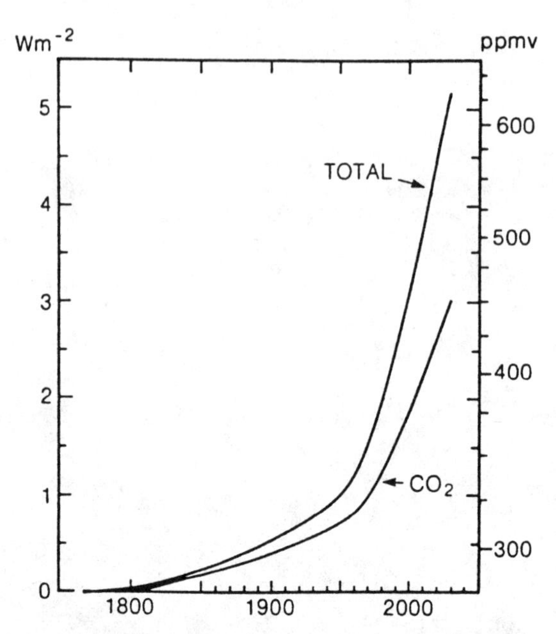

Fig. XII-2. History and future of greenhouse gases in Earth's atmosphere. Sharp rise began in 1950's and is projected to rise steeply in the future. (From T. M. L. Wigley. 1987. Geophysical Research Letters *14(11): 1135–1138.)*

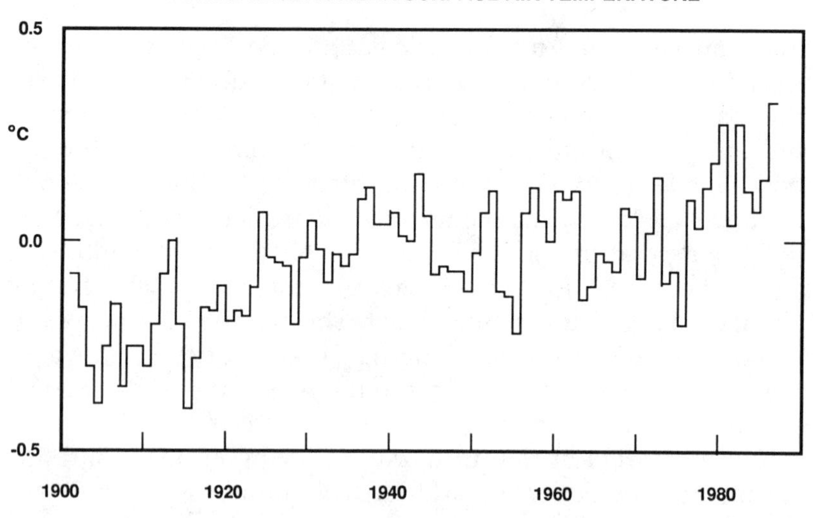

Fig. XII-3. Twentieth century log of global surface air temperature. (Data from P. D. Jones et al. (1988).)

Reckless as our actions may seem, they pale against the dangers lurking if the warming of the climate is not, as expected, gradual over the next century but is instead sudden and abrupt. For then we will not have the opportunity to adapt by continuous, small adjustments. Geologic history offers no comfort: whenever it records major, sharp changes in temperature, it also reveals corresponding sharp changes in plant life. Land areas that supported forests during warm epochs have on occasion been quickly turned into tundra, for instance. If the same should happen to our agricultural crops, the consequences for humanity would be dire indeed. And who is to say that is not exactly what will happen?

As another example of the magnitude of human modification of the biosphere, note that the oceans already receive significantly more metals from byproducts of man's activity than they do from natural events. Or compare recent human activities

Table XII-1. Quantities of Metals Entering Oceans[a] *(In Thousands of Tons Per Year)*

Metal	Natural Processes	Human Processes
Iron	25,000	319,000
Manganese	440	1,600
Copper	375	4,460
Zinc	370	3,930
Nickel	300	358
Lead	180	2,330
Molybdenum	13	57
Silver	5	7
Mercury	3	7
Tin	1.5	166
Antimony	1.3	40

[a] Data from G. O. Barney, Director, *The Global 2000 Report to the President*, Washington, D.C.: U.S. Government Printing Office (1980).

with previous cosmic catastrophes, for the two are indeed comparable in magnitude. Humankind's environmental modifications are now so massive that other species are going extinct at a rate seen before only at times when giant meteorites (or asteroids or swarms of comets) bombarded the earth. During the next thirty years, barring an unforeseen change in human lifestyles, a million species could be lost—forever! And the billion people so recklessly experimenting with Gaia's resiliency are a mere one-seventh of the population expected by the year 2000. What will happen if their brethren in the underdeveloped world demand an equivalent material lifestyle?

Burgeoning population is one worry; its reduction to zero another. The means for accomplishing the latter are as omnipresent as the "nuclear buttons" constantly at the sides of two individuals leading nations with opposing ideologies. If neither can accept the existence of the other, then both will perish—and with them the rest of humanity.

We have reached a critical bottleneck separating past from future. But there is no reason why it cannot be navigated. the enormous knowledge base that humanity has accumulated contains strategies for survival—if only humans will accept the new responsibilities thrust upon them. In other words, the fact that we can *foresee* the outcome of certain actions demands that we do so. If we want life to continue into the future, we can ensure that it will. We can calculate—at the least, estimate—the future consequences of various optional actions taken now, then choose the option most favorable for ensuring life in the future. If we do so, the *natural selection* that led to our existence will be joined by *conscious selection* directed by us. Changes of this significance in the way the universe evolves have occurred only a few times in all of cosmic history. And we are here at one of those major transitions, guiding its outcome. We matter indeed! We can ensure *immortality*, a goal for which humans have striven ever since we discovered our finite lifetimes, and which is now within reach provided we define it as an eternal superorganism, not as an ageless individual.

The fact that biological evolution via natural selection is about to be supplemented with a directed mechanism for change does not mean natural selection contains no useful lessons. The first lesson—one we must grasp immediately—is that nature is an inexhaustible resource, a storehouse of the possible, a repository of novelty. The information pool nature has accumulated over billions of years through innumerable recombinations of matter in energy-rich environments has scarcely been tapped during the mere tens of thousands of years that humans have devoted to its study. We advance not at all if every increment we contribute to an information pool is matched by an equivalent deletion from it by a human-caused extinction.

Humans must also learn to balance two seemingly competing demands. On the one hand, our participation in a superorganism requires that we behave in a cooperative, coordinated manner with all other peoples of the world, with all other kinds of living things, and with all the matter that is presently inanimate but will someday constitute portions of animated creatures. On the other hand, the strength of life's ability to adapt, hence survive, has always been its *diversity*. Since the key to the new type of evolution we embark upon is knowledge and information, we need likewise to keep those resources rich and diverse. The cooperation among peoples of the world necessary for their mutual survival must not be achieved at the expense of homogenizing every language, every cultural tradition, every politico-economic system, or every life style. Individuality must be preserved within a global community. The reservoir of knowledge must contain enough variants to cope with problems we cannot yet imagine.

The surest way to enrich the knowledge pool that will keep the flywheel of cultural evolution turning is to nourish the human spirit of *curiosity*. Our natural wonder about the universe must compel us to explore it aggressively. Our understanding must never be curtailed by lack of effort to advance it. Every slackening of effort represents diminished options in the future. It also limits human creative potential at present, denying man full use of the one property distinguishing him most clearly from the other inhabitants of the superorganism. It restricts his opportunities for the future and denies him the possibility of excellence.

Humans themselves contribute strongly to their potential stagnation by increasing their numbers. As people are forced into closer contact, they must submit to greater regimentation. Each loses freedom because his actions affect not only himself, but also his neighbors. But regimentation leads to a collapsing of ideas, a constriction of opportunity, and eventually an end to cultural evolution; for evolution has always progressed through disequilibrium. If all must (or do) "march to the beat of the same drummer," a stable state will be achieved—and it will be a terminal one.

We have defined life as a mysterious mechanism for extracting negentropy (order) from a "flow" driven by the existence of various disequilibria. The details of the mechanism's operation are intricate, but unimportant. In this broadest cosmic framework, any system that creates order from chaos can be termed *alive*.

Our universe began in chaos. Yet it changed, beginning almost immediately after its formation, into the hierarchy of organized structures populating it today, including the most ordered, the living. Today's conditions derive from initial ones that could have varied very little if life were to arise some time after the birth of the universe. Birth is exactly the proper word to describe its origin, since the universe fulfills our definition of life as an order-creating system. In other words, *the universe is living,* and has been since its inception. It—meaning all space considered for all time—is the unit constituting the true Gaia. Earth's superorganism is a mere, and perhaps temporary, detail in the operation of this ultimate living system. Any organism, man included, that serves functionally (and again temporarily) within the superorganism is an even less significant particular. But we are not trivia. Since we comprehend what it means to be alive, we are the "sensory organs" with which the living universe monitors its own "physiology", if you will—taking its pulse and measuring its blood pressure. Without us, the universe is "blind." Our "vision" into the future enables the universe to continue to live . . . unless our (momentary, let us hope) myopia strikes it dead.

> As Sarpedon says to Glaucus in the *Iliad,* a hundred thousand fates stand close to us always, which none can flee and none avoid. The complexity of the universe is infinite, and the days of a man's life are threescore years and ten.[2]

For Further Reading

Broecker, Walter S. 1987. Unpleasant surprises in the greenhouse? *Nature* 328: 123–126.

Broecker, Walter S. 1987 October. The biggest chill. *Natural History* 96(10): 74–82.

Campbell, Jeremy. 1982. *Grammatical Man: Information, Entropy, Language, and Life,* pp. 254–273. New York: Simon and Schuster, Inc.

Chaisson, Eric. 1987. *The Life Era.* New York: The Atlantic Monthly Press.

Dyson, Freeman. 1979. *Disturbing the Universe.* New York: Harper and Row, Publishers.

Ehrlich, Paul R.; Harte, John; Harwell, Mark A.; Raven, Peter H.; Sagan, Carl; Woodwell, George M.; Berry, Joseph; Ayensu, Edward S.; Ehrlich, Anne H.; Eisner, Thomas; Gould, Stephen J.; Grover, Herbert D.; Herrera, Rafael; May, Robert M.; Mayr, Ernst; McKay, Christopher P.; Mooney, Harold A.; Myers, Norman; Pimentel, David; and Teal, John M. 1983. Long-term biological consequences of nuclear war. *Science* 222: 1293–1300.

[2] A. E. Housman, in his 1892 *Introductory Lecture* to the united Faculties of University College, London. As quoted by John Miles in *Engineering & Science XLVII* (1): 12 (1983).

Fyfe, W. S. 1981. The environmental crisis: quantifying geosphere interactions. *Science* 213: 105–110.
Hamburg, David A. 1984. Science and technology in a world transformed. *Science* 224: 943–946.
Jones, P. D., Wigley, T. M. L., Folland, C. K., Parker, D. E., Angell, J. K., Lebedeff, S., and Hansen, J. E. 1988. Evidence for global warming in the past decade. *Nature* 332: 790.
Seielstad, George A. 1983. *Cosmic Ecology: The View from the Outside In*. Berkeley: The University of California Press.
Turco, R. P.; Toon, O. B.; Ackerman, T. P.; Pollack, J. B.; and Sagan, Carl. 1983. Nuclear winter: global consequences of multiple nuclear explosions. *Science* 222: 1283–1292.

Index